"It's got everything . . . sex, violence, and a cast of millions!"

"Grin and Bear It," from the Vancouver Sun
Courtesy Publishers-Hall Syndicate

Soil Biochemistry

Volume 3

BOOKS IN SOILS AND THE ENVIRONMENT

edited by
A. Douglas McLaren

College of Natural Resources
University of California
Berkeley, California

Soil Biochemistry, Volume 1, edited by A. D. McLaren and G. H. Peterson
Soil Biochemistry, Volume 2, edited by A. D. McLaren and J. Skujiņš
Soil Biochemistry, Volume 3, edited by E. A. Paul and A. D. McLaren
Soil Biochemistry, Volume 4, edited by E. A. Paul and A. D. McLaren

Also in the Series

Organic Chemicals in the Soil Environment, Volume 1, edited by C. A. I. Goring and J. W. Hamaker

Organic Chemicals in the Soil Environment, Volume 2, edited by C. A. I. Goring and J. W. Hamaker

Humic Substances in the Environment, by M. Schnitzer and S. U. Khan

Microbial Life in the Soil: An Introduction, by T. Hattori

Additional Volumes in Preparation

SOIL BIOCHEMISTRY

EDITED BY

E. A. Paul

Department of Soil Science
University of Saskatchewan
Saskatoon, Saskatchewan, Canada

A. Douglas McLaren

College of Natural Resources
University of California
Berkeley, California

Volume 3

MARCEL DEKKER, INC., New York

MARCEL DEKKER, INC.
270 Madison Avenue, New York, New York 10016

LIBRARY OF CONGRESS CATALOG CARD NUMBER: 66-27705

ISBN: 0-8247-6140-5

Current printing (last digit):
10 9 8 7 6 5 4 3 2 1

PRINTED IN THE UNITED STATES OF AMERICA

PREFACE

Although the study of organic components and enzymatic reactions in soil has a long history, biochemical aspects have received emphasis and recognition only recently. Volume 1 of Soil Biochemistry, published in 1967, stresses the occurrence of organic compounds and enzymatic systems present in soil. The second volume includes discussions of new techniques, biochemical transformations and the colloidal aspects of humus and clays.

Scientific progress is dependent on the emergence of new scientific concepts with an expanding vista. In this context, terrestrial soil is a subsystem involving the interaction of stabilized enzymes, living organisms, accumulated organic matter, and colloidal inorganic constituents. In this volume our authors have analysed some of these subsystems qualitatively and have indicated how quantitative methods draw attention to areas of scanty or missing information and a paucity of concepts. With mathematical modeling procedures, one is forced ultimately to give meaningful content to what would otherwise be only accommodating and arbitrary parameters. Only then can the models have predictive value in real systems. Some of the methodology and principles governing stream beds, estuaries, oceans, and sewage sludge systems are also useful in the approaches to soil biochemistry. Thus it behooves the soil biochemist not to work in isolation. Team research and a strong feeling for the applicability of data from what used to be considered as widely divergent fields probably best characterize today's soil biochemist interested in environmental problems. This book surveys this notion from many directions.

E. A. Paul

A. D. McLaren

CONTRIBUTORS TO THIS VOLUME

P. A. CAWSE, Environmental and Medical Sciences Division, Atomic Energy Research Establishment, Harwell, Didcot, Berkshire, England

C. H. DICKINSON, Department of Botany, University of Newcastle-upon-Tyne, England

P. H. GIVEN, Fuel Science Section, College of Earth and Mineral Sciences, The Pennsylvania State University, University Park, Pennsylvania

A. D. McLAREN, Department of Soils and Plant Nutrition, University of California, Berkeley, California

E. A. PAUL, Department of Soil Science, University of Saskatchewan, Saskatoon, Saskatchewan, Canada

J. S. RUSSELL, CSIRO, Division of Tropical Agronomy, Cunningham Laboratory, St. Lucia, Queensland, Australia

GEORGE H. WAGNER, Department of Agronomy, University of Missouri, Columbia, Missouri

TOMIO YOSHIDA, The International Rice Research Institute, Los Baños, Laguna, Philippines

CONTENTS

George H. Wagner

CONTENTS OF OTHER VOLUMES

CHAPTER 1

BIOCHEMISTRY OF THE SOIL SUBSYSTEM

E. A. Paul

Department of Soil Science
University of Saskatchewan
Saskatoon, Saskatchewan, Canada

A. D. McLaren

Department of Soils and Plant Nutrition
University of California, Berkeley

I. INTRODUCTION

The field of soil biochemistry traditionally has been
discussed under the headings of plant-microbial interactions, soil-
microbial relationships, and soil enzymes. Special emphasis has
often been placed on the role of these interactions relative to the
fertility status of cultivated soils. An aspect of soil biochemistry that is gaining increasing significance is a description of the

1

role of soil organisms and their enzymes in terms of ecosystem concepts. The soil, with its organisms and its ambient, can be designated an ecosystem, especially under fallow or laboratory conitions. However, the main function of the term "ecosystem" in ecological thought is to emphasize the relationships between organisms and their environment, especially the interaction between heterotrophs and autotrophs [1]. In this context the soil as discussed in this chapter is best referred to as a subsystem, as indicated in the title. Microorganisms, and their enzymes, often in conjunction with the soil fauna, bear the burden of energy and nutrient transfer through the various decomposition processes. In addition, the biomass of organisms and their metabolites constitute a significant portion of fixed energy and nutrients in the system, making it necessary to estimate their biomass and nutrient content [2-4].

The interaction of plants, especially the root system, with the microorganisms and fauna of the soil is receiving a great deal of attention in today's scientific literature. The effect of soil physical and chemical characteristics and the influence of the environment are also being stressed with emphasis on quantification of data. The following brief discussion suggests how some of the traditional problems can be considered from the systems analysis viewpoint with a short list of references concerning the biological interactions in the soil subsystem. Some of the more recent reviews describing the concepts involved are cited to provide access to a detailed literature. The principles of investigation and much of the data concerning the biochemistry of soil are applicable to a wide range of terrestrial ecosystems, and soil systems also have a great deal in common with aquatic and sewage systems [5, 6].

II. PLANT-MICROBE RELATIONSHIPS

Authors of botany texts have traditionally presented results of studies of plant physiology and anatomy as if plants were axenic

entities. However, there is now abundant evidence that microbes
that are not considered pathogens can affect root morphology and
nutrient uptake as well as the metabolism of the entire plant [7,
8]. Microbes growing in the rhizosphere are not as effective as
those in the rhizoplane, including those in the mucigel [9], and
continuing research must take into account some observations of
Darbyshire and Greaves [10]. They found an interaction between a
Pseudomonas sp. and the soil amoeba *Acanthamoeba*. The amoebae
alone probably can only colonize the root surfaces of peas, although
mucigel can be entered. Amoebae accompanied by bacteria can be
observed inside root cells if both are present during root growth,
and it is probable that the amoeba follow the bacterial invaders.
When peas were grown in garden soil sterilized by γ-irradiation and
inoculated with *Pseudomonas* sp. bacterial invasion also occurred;
this was not observed in unsterilized inoculated soil. The results
suggest the possibility that "normal" soil-plant-microbe relationships may extend all the way from the rhizosphere to the cortex.

A.   Effects on Plants

Even with a reduction in root hair growth in the presence of
rhizoplane organisms, nonsterile roots can take up more phosphate
than sterile roots. Bowen and Rovira [8, 11], using short pulses
of ^{32}P in solution concentrations equivalent to those found in
soil solution, showed that nonsterile plants had both greater uptake
and greater translocation of phosphorus. Miller and Chau [12], comparing plant growth in sterilized soil with that in unsterilized
soil in plastic film isolators, found that the presence of microorganisms was accompanied by an increased mass of tops and roots as
well as a higher plant content of Ca, Mg, Fe, Al, Cu, Zn, and Mo.
Total plant contents of N, P, and K were not affected by microbial
activity. Other uptake studies have shown varying results [10].
Interception and trapping of nutrients by microorganisms on roots
[13, 14] have been noted especially under conditions of nutrient
deficiency.

In one study of plant-microbe interactions, 13 separate strains (including short gram-negative rods, gram-positive spore-forming rods, and *Streptomyces* sp.) were isolated from root surfaces of tomatoes grown in Columbia soil by streaking on a soil extract agar (H. A. Habish and A. D. McLaren, 1964). Only one of these, a gram-positive rod, reduced seed germination on moist filter paper. Shoot development was less sensitive to inhibition by bacteria than was root development. On agar slants containing Hoagland's solution, in light, roots could be colonized by all the strains but root growth inhibition was observed with only four of the cultures; one inhibited lateral root development, one inhibited the main root, and the other two showed a general inhibitory affect. By contrast, an inoculum of nonsterile soil did not inhibit root development, showing that roots were unaffected by a mixed population.

Inoculation of five of the 13 strains separately into electron-sterilized soil in which tomato seedlings with sterile roots were growing with Hoagland's nutrients [15] showed no effect compared with plants in sterile soil after 8 weeks of growth. Even a culture that was an active inhibitor of root development on agar failed to affect shoot growth under these conditions. When nutrients were not added to the sterile soil, one culture reduced the fresh weight of seedlings by half. Inoculation with a mixed population of microbes provided by nonsterile soil almost doubled the plant weight. To understand the plant-microbe interactions in the rhizosphere we must know by what mechanism microbes bring about these differences and whether these mechanisms are important under natural growing conditions. Plant hormones and growth inhibitors are synthesized by many soil bacteria and these must be involved [16].

Previous conjectures that roots of plants growing in association with microorganisms have a tendency to contain higher concentrations of certain amino acids than sterile roots [7] have been confirmed (R. Prasad and A. D. McLaren, 1968). Whether microorganisms secrete amino acids that are taken up by roots or whether they secrete inhibitors of protein synthesis and thereby cause an accumulation of free amino acids is unknown.

Humic compounds, either of microbial or of phenolic plant degradation origin, have long been postulated to affect plant growth [17]. Working under sterile conditions, Flaig [18] has recently measured the uptake and transport of ^{14}C -labeled phenol carboxylic acids and substituted hydroquinones by growing plants. These compounds, or their transformed products, were shown to have both direct and indirect effects on plant growth. The phenolic compounds were transformed into glucose esters and glucosides with some oxidative decarboxylation occurring after uptake.

B. Effects on Microorganisms

Many workers have shown that the activity of nitrifying bacteria is inhibited by exposure to either intact living plant roots or aqueous extracts of these roots [19,]0] and the concept that climax vegetation tends to produce nitrification inhibitors [21] is interesting and merits consideration by botanists and plant ecologists. Other workers have expressed doubt that inhibition of nitrification is caused by toxins. Soil acidity [22] and low populations of nitrifiers [23] also have been suggested as possible explanations for the low nitrification rates. However, Huntjens [24, 25] (using ^{15}N) has attributed the accumulation of organic nitrogen and the low net nitrogen mineralization rates in permanent pastures to root excretions and to dead roots, which supply carbon for immobilization of mineralized nitrogen into microbial tissues, thereby leaving only negligible amounts of NH_4^+-N in the soil. Knowledge of heterotrophic nitrification [26, 27] in nature would affect conclusions drawn from work on autotrophic nitrifying bacteria but would not alter concepts based on those low quantities of mineral N because of a balance between mineralization and uptake by plants.

Changes in the selective action of root exudates exert a differential influence on rhizosphere microbial interactions [28]. Elkan [29], working with nodulating and genetically related non-nodulating soybean lines, reported differences in total numbers and

nutritional requirements of isolates from the rhizospheres of the
two types of plants. Mytton and Gareth Jones [30] have presented
evidence showing that the total volume of white clover nodules can
be increased by selection of the host plant. However, after the
first generation of selection the increased nodule volume is not
accompanied by an increase in the plant growth. Another example of
plant control of the rhizosphere flora is shown in relation to root
rot of wheat [31]. Substitution of a chromosome pair from a variety
of root-rot resistant spring wheat into a highly susceptible variety
altered both the resistance to root rot and rhizosphere microflora
of the new plant qualitatively and quantitatively. A larger percentage of the rhizosphere isolates from the resistant parent and
the substitution line were found antagonistic to the root-rot
organism (*Cochliobolus sativus*). This further indicates that the
rhizosphere population can be controlled by genetic manipilation of
the host plant.

Specific aspects of the occurrence of N-fixing members of the
family Azotobacteraceae in tropical and subtropical environments
have been stressed. High *Azotobacter* populations have been found in
the soils of Egypt [32]. Certain forage grasses, rice, and especially sugar cane greatly stimulate the multiplication of
Beijerinckia and depress the number of amino acid requiring bacteria
and actinomycetes [33]. A plant-bacteria association between
Azotobacter paspali and the grass *Paspalum notatum* has been noted
[34]. Nitrogenase activity, measured by C_2H_2 reduction in the
rhizosphere of the cultivar Batatais, produced 1-32 nmoles C_2H_4 per
gram dry weight per hour, whereas a cultivar not colonized by *A.
paspali* when exposed to C_2H_2 produced less than 0.5 nmoles C_2H_4/g
per hour. Activity of soil cores containing plants with leaves attached was little affected by pO_2 changes and showed no diurnal
fluctuation in activity during a 16-hr day/8-hr night period. Nitrogen fixation by the association was estimated to range up to 90
kg/ha per annum [35]. The occurrence of nitrogen fixing *Klebsiella*

in the rhizosphere of leguminous plants has also been noted [36]
but the significance of this relationship is not known.

Macura [37] in a recent review stated that: "The process of
colonization of roots by microorganisms and the function of control
of the rhizosphere population are so closely associated that none
of the individual processes can be fully elucidated without clari-
fying at least the basic features of the other." The availability
of adequate methods of soil sterilization, tracer techniques, lami-
nar flow cabinets, plastic film isolaters, and plants that can be
genetically manipulated to alter the rhizosphere microflora should
make possible the separation of the different rhizosphere effects,
at least for a specific set of conditions. Because of the complexity of the interactions it will probably continue to be difficult
to make generalizations.

C. Effects on Soil Respiration

Rhizosphere soil is generally considered to have higher
respiration rates than nonrhizosphere soil. Similarly, nonsterile
roots are said to respire more than sterile roots. Some soils,
such as grasslands, contain so many roots that they are often considered to be completely rhizosphere in nature [38, 39]. However,
the impact of the rhizosphere probably has to be interpreted in a
different manner when one considers, and models, energy transformations of the soil subsystem. Katznelson and Rouatt [40] found a
threefold higher respiration rate for rhizosphere than for nonrhizosphere soils, but quantitative estimates of the amount of
rhizosphere soil present were not made. The classical method of
determining root respiration has been based on a comparison of CO_2
respired from cropped and from fallow or clipped plots. This has
led to the general conclusion that plant roots account for 30-50%
of the soil respiration [41]. Extrapolation of more recent data,
obtained with an infrared gas analyzer, to a soil free of roots
results in similar values [43]. The use of short-term labeling

with ^{14}C results in lower estimates of the contribution of
roots to the total respiration of a virgin grassland [43].

The microflora of the phyllosphere have been calculated to
contribute a negligible portion of the microbial biomass of a grassland ecosystem [20]. Somewhat similar conclusions can be drawn
from the data for root-associated bacteria if plant count data are
used as a measure of actively metabolizing organisms. The numbers
of bacteria on the rhizoplane and the closely associated soil will
usually have a rhizosphere-to-soil ratio of 25:1 [8, 37]. However,
even in a grassland soil, roots constitute less than 1/300 of the
total weight of the soil, indicating that plant-associated organisms
comprise less than one-tenth of the viable population of bacteria.
Determination of the biomass of the bacteria by microscopic measurements would probably indicate a still lower percentage of the total
biomass of microorganisms immediately associated with the plants.

Schappert [44] has shown that the respiration of a fallow soil
can be higher than that of a nearby grassland when calculated over
a complete growing season. This was attributed to better moisture
conditions in the fallow, leading to higher microbial respiration
rates than the combined root and soil organism contribution on the
native grassland. Although the rhizosphere population can have a
marked influence on plant growth as discussed above, its significance when considered in the concept of ecosystem energy flow via
the decomposer cycle may require a different interpretation.

III. SOIL-MICROBE RELATIONSHIPS

Observations of micro- [28] and molecular-environmental [45]
influences on microbial activity are necessary for an understanding
of the role of abiotic controlling factors [46]. One of the most
interesting observations is the result of a study of streptomycetes,
which occur in an acidic podzol with a bulk pH below that which
permits growth in pure culture. Williams and Mayfield [47] suggest
that occasional periods of activity by neutrophilic *Streptomyces*

occurring in acid soil are found in microsites containing adsorbed
ammonia. They showed, with the aid of color pH-sensitive indicators
and microscopic observations, that the pH of organic particles exposed to nitrogenous metabolites could be as much as 2 pH units
higher than that of the surrounding soil. Observations on unamended
soil revealed no sites with a pH higher than that of bulk soil.
This situation and the later observations [48] that periods of active growth are discontinuous in space and time are in keeping with
a general discussion of surface pH theory in Chapter 1, Volume 1,
of *Soil Biochemistry*.

A. Experimental Models

The study of environmental influences on the soil subsystem
brings forward the contrast between "closed" and "open" systems.
In both kinds of systems the concentration of a metabolite(s) is a
function of time t and the rate of change is a function of the
amount of catalyst E, the temperature, the pressure, the water activity, amounts of coreactants, surface pH, bulk pH, and the
numbers of cells and species of the active cell population. In an
otherwise closed system, if a substrate at some bulk concentration
[S] is added to a sample of moist soil (substrates might be H_2S,
$S_2O_4^{2-}$, NH_4^+, or NO_2^-), the substrate becomes distributed among a
number of surface and colloidal sites by adsorption and by ion exchange; therefore, the concentration can no longer be accurately
specified. For this reason, any model of the chemical activities
of soil is at once, of necessity, a gross oversimplification. In
an isolated soil system, whether open to air or not, the organisms
are distributed in the system as zooglea, in pairs, singly, as
spores, etc., both in easily elutable and in tightly bound forms on
the mineral and humus particles (K. C. Marshall, *Soil Biochemistry*,
Volume 2). The products P (e.g., SO_4^{2-} or NO_3^-) of reaction, secreted by any microbe or mediated by extracellular enzyme action,
are mixed more or less uniformly and are available by diffusion
and/or by microbial movement for further chemical reaction. On a

macroscale, for example, the ratios $[NH_4^+]/[NO_2^-]/[NO_3^-]$ are constant throughout the bulk sample of soil. Customarily one expresses a rate of change of S in such a closed system as $-d[S]/dt = d[P]/dt$; these are scalar quantities; i.e., they represent changes in concentration with respect to time only and no translocation in macrospace is involved.

In nature, however, soils have a profile, i.e., a change of properties with depth. If NH_4^+, as substrate, is supplied at the surface of the soil and a solution with concentration $[NH_4^+]$ moves downward, it can be oxidized by *Nitrosomonas* sp. to NO_2^-. If $[NO_2^-]$ is small initially at the surface, *Nitrobacter* sp. will be exposed to only small concentrations of NO_2^-, whereas at some depth below the surface, X, the concentration of NO_2^- will be greater and *Nitrobacter* will experience an increased availability of nutrient. The concentrations and ratios of concentrations of NH_4^+, NO_2^-, and NO_3^- will not only vary with time in a given volume element but, in contrast with a closed, uniformly mixed system, will vary from volume element to volume element with a downward change in X. An organism, such as *Nitrobacter*, situated lower in the soil profile will be exposed to ratios of $[NH_4^+]/[NO_2^-]/[NO_3^-]$ differing from those in volume elements above it. The best way to express the changes of bulk concentrations in the system is as $d[NH_4^+]/dX$, $d[NO_2^-]/dX$, and $d[NO_3^-]/dX$. (By "bulk concentrations" are meant the total amounts of each of S_i divided by the bulk volume element of the soil at any X.) These are clearly vector quantities because they have both magnitude and direction (downward in this case). Therefore, soil biochemistry has the character of vector chemistry in common with other transport situations in nature (cf. Chapter 1, Volume 2 of *Soil Biochemistry*. Uptake of sugars by plant roots or transport into intestinal villi are other examples [49].

An interesting contrast between open and closed systems is as follows: If Hanford fine sandy loam is reperfused with NH_4^+-N, nitrification takes place readily, whereas the same soil in a column with continuous passage of substrate shows no nitrification unless

micronutrients are added to the ammonium solution (L. Dinkins and
L. Belser, (1970). Evidently with reperfusion of soil in the same
closed system, traces of micronutrients originally in the soil are
retained, whereas under continuous addition of new substrate such
nutrients are quickly lost from the soil by ion exchange with
NH_4^+, followed by elution from the column.

The question of microbial growth during reperfusion is not
easy to handle mathematically. Nishio and Furusaka [50] have made
a careful study of the kinetics and numbers of nitrite-oxidizing
microbes during reperfusion of nitrite. They observed an increase
in numbers of nitrifiers during repeated reperfusion with fresh
nutrient, even though rates of nitrification became seemingly constant. L. Dinkins and L. Belser, (1970) find that the constancy of
nitrification under similar conditions is not caused by limited
solubility (or limiting rate of solution) of oxygen and that the
numbers do become constant if reperfusion is continued long enough.

A number of workers have been studying mathematical models for
biochemical reactions in flowing systems, which take into account
growth of organisms [51-54] and hydrodynamic dispersion [55]. Cho
has shown how concentrations of exit solutions may be corrected for
ion exchange and hydrodynamic dispersion (Fig. 1). In practice,
dispersion may not be as important a correction as is loss from denitrification [56].

A simple model will now be described mathematically; a more
complicated model will be discussed in Section V, C.

Let us assume that the rate of oxidation of NH_4^+ or NO_2^- is given
by

$$- d[S]/dt = (A + B)\gamma m + \alpha m + k'm \qquad (1)$$

where m is oxidizer biomass, A is the nitrogen oxidized for energy
for cell growth per unit weight of biomass synthesized, B is the
amount of substrate nitrogen incorporated in cells per unit weight
of biomass synthesized, γ is the specific growth rate constant, α
is the N oxidized per unit weight and time for maintenance, and k'

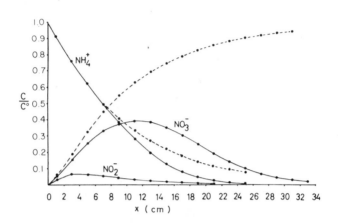

FIG. 1. Variations of concentrations, proportional to C, with depth and with (solid lines) and without (dashed lines) hydrodynamic dispersion during nitrification. In either situation, curves for nitrite are nearly the same. This figure may be compared with Figure 1-1, uncorrected for dispersion, in Volume 2 in this series. Time of flow, 200 hr; velocity, 0.1 cm/hr. Data courtesy of C. M. Cho [55].

is the N oxidized in ways that do not lead to growth or maintenance. Now γ = (1/m) (dm/dt) and the first term can be ignored under two limiting conditions. If the population reaches a maximum, dm/dt = 0 and a steady state pertains. Because γ is quite small for nitrifiers and can correspond to division times of as long as 36 hr in soil, dm/dt can be taken as zero if the time of the experiment is short by comparison. In either case the A, α, and k' terms represent waste metabolism and the products, NO_2^- or NO_3^- are waste.

It has recently been found that populations do reach maxima, independent of depth in a soil (diluted with sand) column after about 3 weeks of continuous flow [57]. Combining the last two terms in Eq. (1), and assuming Monod relationships among the constants and the substrate concentration [57], we obtain Eq. (2).

$$- d[NO_2^-]/dt = k_1 m[NO_2^-]/(K_m + [NO_2^-]) \qquad (2)$$

Here k_1 is a sum of constants and K_m is a saturation constant for substrate with the enzyme system involved. For piston flow $X = f \cdot t$, where f is the flow rate in the column (Chapter 1, Volume 2 of this series) and t is flow time. Substituting for time in Eq. (2) and integrating we obtain

$$[NO_2^-] = K_m \ln ([NO_2^-]_0/[NO_2^-]) + [NO_2^-]_0 - KX \qquad (3)$$

for the concentration of nitrite solution flowing within the column at a depth of X. $K = k_1 m_{max}/f$, where m_{max} is the maximum population of nitrifiers. Figure 2 shows how nitrite varies with depth, at two

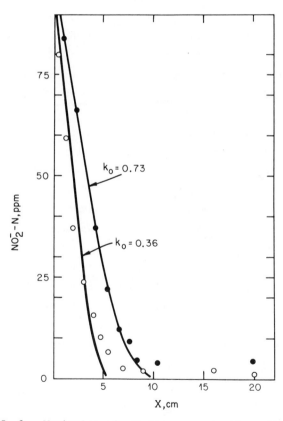

FIG. 2. Variation of nitrite concentration with depth in a column containing 10^5 nitrifiers per cm^3 of sandy soil at two entering flow rates (k_0) of 0.36 and 0.73 cm/hr. The lines are calculated with the aid of Eq. (3) and $K_m = 16$ ppm NO_2^--N. (Unpublished data of J. Rehboch, S. Ardakani, and A. D. McLaren.)

flow rates, in a column of 90% sand mixed with 10% Hanford fine
sandy loam. The model seems to be satisfactory as a first approximation.

B. Field Interpretations

From this model, in which soil was diluted with sand, we can
predict that almost complete nitrification of added nitrite can take
place in the first centimeter or two in Hanford soil in the field.
This has been found and is consistent with the observation that
rarely do plant roots suffer from nitrite toxicity in natural set-
tings. During nitrification, populations of other soil organisms
also tend to change [58]. Soil as a chromatographic column can
change the ^{15}N / ^{14}N ratio during translocation. Nitrogen fixers
and oxidizers including *Nitrosomonas* sp. may discriminate in favor
of ^{14}N. This will be of interest in any vector analyses that
depend on these ratios [59].

The ^{15}N / ^{14}N ratios of nitrate nitrogen in surface waters
are being used to speculate on the relative contributions of fertilizer and soil N [60]. Compared to air, soil is generally enriched
in ^{15}N [61]. The relative enrichment of ^{15}N was used to
calculate the possible contribution of fertilizer N to the ground
water emanating from drainage tiles, and it was concluded that
fertilizer nitrogen contributed 55 ± 10% of the NO_3^--N entering the
lake under study during the spring months of 1970. This study is
another example of vector analysis, similar to that discussed above,
but has been attempted on a much larger scale. Natural variations
in ^{15}N in soil and water [61] and discrimination against the
heavier isotope during movement through the soil profile, however,
must be considered in interpreting such data [62].

IV. SOIL ENZYMES

The subject of soil enzymes has been reviewed in Volume 1 of
this series, and in Volume 2 McLaren and Skujins considered indirect
evidence for extracellular phosphatase. [N.B.: On p. 9, Volume 2,

lines 6-9 should read: "In Fig. 1-2 it can be seen that the
variation in activity of phosphatase with the season is much less
than the fluctuation of microbial populations...".] By extraction
of soil with dilute buffer, small amounts of cell-free enzymes have
been obtained as sols or solutions; these include urease [63],
uricase [64], and peroxidase [65]; and an enzyme that degrades
Malathion has been partially purified from cultivated fields of
western Washington [66]. Generally the fraction of total enzyme
activity of soil represented by these soluble extracts has not
been noted.

Chalvignac and Mayaudon [67] have extracted a beechwood soil
with 0.2 M NaHCO$_3$ and obtained (after neutralization) a sol with
tryptophan decarboxylase activity. The neutralized sol did not
sediment in a field of 20,000g for 30 min and was brown in color.
It lost 60% of its activity during heating at 100°C for 10 min and
maintained 74% activity after lyophilization. The sol retained
half its activity if subjected to conditions of the classic humic
acid preparation, namely exposure to 0.1 N NaOH followed by precipitation at pH 2 with HCl. The sol also was active with or without
the presence of oxygen. Burns et al. [68] reported the extraction
of a urease-active organocomplex from a Dublin clay loam with a
mixture of urea, sodium chloride, EDTA, mercaptoethanol, and
phosphate at a neutral pH. Urea acts as a hydrogen-bond breaking
reagent and, because of the presence of urease in the humus, is a
"self-destructing" extracting agent. The extracted material was
free of clay and microbes. About one-third of the urease activity
of the Dublin soil was accounted for in the extract. The urease
activity of this humus complex was resistant to the proteolytic
enzymatic activity of added pronase.

Urease activity is nearly ubiquitous in soils and its origins
have been reviewed [69]. It is probable that native soil urease
resides in organic colloidal particles with pores large enough for
water, urea, ammonia, and carbon dioxide to pass freely but with
pores small enough to exclude such enzymes as pronase, [70] as
illustrated in Fig. 3. A postulate that there is a protective

□ Substrate ▭ Product
E, enzyme (e.g., urease U) ∿ humus

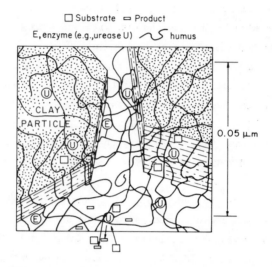

FIG. 3.   This model for soil enzyme location and activity
consists of enzyme embedded in, and perhaps chemically attached to,
a humus polymer network in contact with clay particles.   Substrates,
such as urea, can reach the enzyme by diffusion through pores too
small for enzymes to penetrate. [R. G. Burns, A. H. Patrick and
A. D. McLaren, Soil Sci. Soc. Amer. Proc., 36, 308 (1972).]

action by clays [71] on soil enzyme-protein stability is not required. The kind of attachment between soil humic residues and
enzymes now becomes a problem of importance alongside those of
exchange capacity, redox capacity, clay complexing properties, and
waterholding capacity of soil humic substances. One possible mechanism is that humic acids bind enzymes by amino-carboxyl salt
linkages [72]. Such complexes should more easily dissociate under
laboratory conditions [73] than complexes with covalent bonding (via
condensation, for example) between amino groups of enzymes and
phenolic residues of humus [74-76]. Hydrogen bonding of humic
proteins to humic surfaces as described by Simonart et al. [77] and
Biederbeck and Paul [78] is another means of stabilizing the enzyme.
Ladd and Butler discuss the role of humic-enzyme complexes more
fully in Chapter 5 of Volume 4. Michaelis constants for soil
enzymes are being gathered (e.g., [79]), but these are only apparent
and will require correction for electrostatic, structural, and
hydrodynamic dispersion effects [49].

V. THE DECOMPOSER CYCLE IN SOIL RESEARCH

The ecosystem research initiated in the last few years and the present stress on the environment have required soil microbiologists and chemists to redefine a number of their concepts and priorities. Russell, in Chapter 2 of this volume, defines "ecosystems" as those systems in which there is an interaction between soil physical and chemical factors and organisms, including microorganisms and man. Another definition says that an ecosystem results from the integration of all the living and nonliving factors of the environment for a defined segment of space and time [4]. The use of a common denominator, such as energy flow, in conjunction with measurements of nutrient cycling throughout the system makes it imperative that the significance of a particular reaction being studied be expressed in relation to the other reactions occurring in the system [80]. The necessity to quantify these reactions in terms such that modern computer technology can be applied (cf. Chapter 1 of Volume 2) raises a host of new questions in the field of soil biochemistry.

A. Measurement of Microbial Activities

A large part of the biologically fixed energy of terrestrial soils is found in the soil as organic carbon. This is released at a rate primarily dependent on the oxidation of the C-H bonds in the material and is dependent on the chemical and physical nature of the materials, the moisture content, aeration, temperature, the structure of the soil, and the type of organisms present. Ecosystem models are usually designed to work in energy flow parameters through various compartments. Caloric values for fauna and higher plants are available [81, 82]. However, it is only recently that data for bacteria and fungi [83] have become available. Gorham and Sanger [84] found that the energy content of soil organic matter was distinctly different in a number of habitats, being high in aquatic soils (5.24 kcal/g) and low in swamp soils (4.87 kcal/g). The relative content of undecomposed plant materials and the

proportion of aliphatic side chains to aromatic rings in soil
organic matter influences the caloric value [85]. The residue remaining after 6 N HCl hydrolysis of a grassland soil contained the
highest energy values when expressed either on an organic matter or
a carbon basis. Such studies stress the need for a common base,
such as caloric content, for measurements. The high ash content of
soils and many organic matter fractions makes these measurements
difficult to obtain. Organic carbon is a more familiar measurement
and is a very useful basis for comparison.

The necessity to quantify microbial and faunal activities in a
broad range of habitats has led to the development of a number of
useful techniques. Enzyme assays continue to be utilized. Ladd and
Butler in Chapter 5 of Volume 4 have indicated that many enzymes
capable of acting extracellularly are stabilized in the soil. This
background activity affects the interpretation of many enzyme assays
in soils [86].

Adenosine triphosphate concentrations (ATP) are often taken
as proportional to biomass and appear to be useful for assessing
microbial activity [2]. When ATP activates a luciferin-lucifurase
enzyme system the light emitted is measured by a sensitive photo-
tube. This technique has been successfully applied to aquatic
systems [87] and to terrestrial soils [88]. Adenosine triphosphate
is not a direct measure of microbial activity or of biomass. Its
concentration drops in senescent cells. Strickland [5] concluded,
however, that a carbon-to-ATP ratio of 250 applies for most aquatic
microflora. Ausmus [89] and Lee et al. [90] found that the ATP
content of individual bacterial cells fluctuated from 1×10^{-10} to
40×10^{-10} µg per cell during various phases of growth cycle with an
average of 2×10^{-10} µg per cell. They concluded, however, that
this technique has great potential as a general indicator of microbial biomass in soils and sediments.

Respiration is the most direct method of measuring activity
under aerobic conditions and a number of techniques are now available for field studies. The infrared CO_2 analyzer makes it possible

to determine CO_2 concentrations to an accuracy of ± 1 ppm. This
instrument can be used either with a canopy or with flux techniques.
Temperature and pressure effects and CO_2 variations in the atmosphere must be taken into account [43, 91]. Using the gas chromato-
graph for determining CO_2, de Jong and Schappert [92] have calculated the flux of CO_2 at the soil surface by measuring CO_2 concentrations at various depths in the soil and determining diffusion
rates.

Plant material uniformly labeled with ^{14}C in a growth chamber
is most often used in decomposition studies [93]. However, ^{14}C
also can be used to label plant materials growing in the field.
This can be taken further to measure root production and turnover of
carbon in the root-soil matrix [94]. It has been suggested that
measurement of the time course of $^{14}CO_2$ evolution by the plant-root-
soil organism system after short-term exposure of the photosynthe-
sising plants to labeled carbon makes possible the separation of
root respiration from that of the rest of the soil [95]. This is
based on the premise that the initial $^{14}CO_2$ entering the soil atmosphere will emanate largely from the roots, whereas later $^{14}CO_2$
will be primarily microbial from degradation of root exudates or
root tissue.

Radiorespirometric methods, as discussed by Mayaudon in Volume
2 of *Soil Biochemistry*, can be applied to the measurement of metabolic processes in soil. Determination of the uptake of ^{14}C-
labeled compounds is also applicable to the measurement of the rate
of utilization of individual organic compounds by aquatic [96] or
sediment [97] organisms. Nitrogen-15 has been employed [98, 99] for
plant studies and isotopes of rubidium, cesium, and zinc have been
used to determine sequential feeding rates of soil microbial and
faunal populations in conjunction with nutrient cycling [100, 101,
and Chapter 1, Volume 2, of this series].

B. Microbial Growth Yields

Kinetic constants for microbial growth are being measured with

pure cultures [102, 103] and for more complex systems involving
aerobic growth of microbial populations in municipal wastes [104]
and aquatic systems [5].

Partly because of the diversity of microbial metabolic activity, the details of the behavior of microbial systems relative to
the energy requirements of cell production have not been completely
elucidated; i.e., the relative values of A, α, and k' of Eq. (1) are
generally unknown. Energy requirements for growth, A in Eq. (1) may
be expressed in terms of adenosine triphosphate (ATP), which couples
the energy-yielding reactions of metabolism to the energy-dependent
reactions in bond formation during synthesis of the proteins, nucleic acid, and wall components of microbial cells. Forest and Walker
[105] have summarized the literature pertaining to the major pathways involved in the synthesis of monomers required in cell growth.
For example, the formation of tryptophan from glucose requires the
expenditure of four high-energy bonds and 1 equivalent of $NADH_2$ per
mole of tryptophan produced. Aspartic acid can be synthesized from
glucose at no energy cost but its conversion to asparagine involves
the loss of two high-energy bonds. Amino acids, such as isoleucine,
use up two ATP's in their synthesis. Formation of an amino acid,
such as arginine, from glutamate requires five high-energy bonds,
whereas synthesis involving a complex series of reactions, such as
that of histidine, involves four high-energy bonds in addition to
the cost of glutamine synthesis. However, two reduced NAD equiva-
lents and an equivalent of 5-amino-1-5'-phosphoribosyl-imidazole-4-
carboxamine, which is a precursor of the purine nucleus, are produced [106, 107].

The major compound in purine nucleotide synthesis is ionosine
monophosphate (IMP), whose formation from glucose requires seven
high-energy bonds directly and four for regeneration of intermediates [105]. The conversion of IMP to GMP (guanosine monophosphate)
requires one more bond. Similar energy requirements have been
established for pyrimidine nucleotides [107, 108]. In calculating
the energy requirement for 30 monomers existing in cells it was

concluded that an ATP change of 0.76 moles was required per 100 g cells produced. This indicates the low requirement of energy in the overall transformation of one molecule to another during the growth of heterotrophs. The next requirement for energy after monomer synthesis is that for polymerization. The energy require- ments for polymerization of amino acids, conversion of nucleotide monophosphates to triphosphate, and synthesis of lipid components also have been reviewed [105]. The pathways for cell wall synthe- sis, however, have not been completely elucidated and calculations concerning their production have assumed that the bacteria cell wall consists of mucopeptide only. The data in Table 1 give estimates

TABLE 1

Synthesis of Microbial Cells from Preformed Monomers[a]

Polymer formed	Amount (g/100 g cells)	ATP required (moles/g synthesized)	Total ATP required (moles/ 100 g cells)
Mucopeptide	15	0.014	0.21
Protein	60	0.045^b	2.7
Lipid	6	0.061^c	0.37
RNA	15	0.017^d	0.26
DNA	4	0.021	0.08
			3.62

[a]Reprinted from [105], p. 224, by courtesy of Academic press.
[b]Average molecular weight of amino acid taken as 110.
[c]Lipids assumed to contain C_{20} fatty acids.
[d]Three ATP required to convert nucleic acid base to mononucleo- tide with an average molecular weight of 300.
[e]Four ATP required to convert nucleic acid base to deoxymono- nucleotide with an average molecular weight of 280.

of the general composition of cells based on the data cited by Mandelstam and McQuillam [107]. The ATP requirement (moles per 100 g of cells) is utilized primarily for protein synthesis, which

includes the majority of the cell constituents. The production of
polymers for 100 g of cells therefore requires nearly four times as
much ATP as does production of the different monomers from a preformed substrate, such as glucose [105].

Morowitz [109] also has given a general, simplified analysis of
bacterial cells. He has assumed that all ribose and deoxyribose was
present as nucleic acids and lipid as phospholipids. Hexoses were
separated into mucopeptide and polysaccharide. However, calculations of ATP requirements for cell material synthesis on this basis
gave nearly identical results to those in Table 1. A total of
0.036-0.037 moles ATP were found to be required for the synthesis of
1 g bacterial cells from glucose. Similar calculations should be
possible for the other major soil microbial component, the fungi, on
the basis of their cell composition and our knowledge of the energy
requirements for production and synthesis of the polymers involved.
This approach, although difficult to apply to such a complex system
as soil, at this state of knowledge should be very useful in helping
to set up experiments that can determine the values involved in a
multicomponent system.

Another approach to microbial growth-efficiency calculations
utilizes total cell yields determined under conditions of active
growth where substrate is limiting. Bauchop and Elsden, in 1960
[110], proposed that the growth of microorganisms is proportional to
the ATP available from substrate degradation. They compared the
yields of different organisms on the basis of the yield coefficient,
Y_{ATP}, defined as number of grams of dry weight of cells produced per
mole of ATP generated from fermentative catabolism. This method is
limited in predictive value, especially for aerobic growth, because
it requires a determination of ATP yields, which are difficult to
obtain under aerobic conditions [105]. This has led Payne and his
co-workers [111, 112] to relate the bacterial biomass produced to
the available electron content of the substrate (Y_{ave-}). The available electron content expresses the degree of reduction of an
organic compound. Because four electrons are required to reduce 1

mole of oxygen, the available electron content of a substrate is
obtained by multiplying the number of moles of oxygen required for
the complete oxidation of a substrate by four. Therefore, glucose
has an available electron content of 24. The yield of cells (ash-
free dry weight) per mole of available electrons was found to average 3.14 for a number of experiments in which growth rates were
limited by substrate availability. Thus, substrates could be directly compared relative to their energy content.

Corrections of the relationship $Y_{ave}-$ have led to expressions
that base the yield on the carbon content of the substrate, expressed as Y_c, the grams of cells produced per gram of carbon in the
substrate [113]. A constant yield coefficient, $Y_c = 1.1$, was found
for sugars, polyhydric alcohols, and paraffins. Carboxylic acids
gave a lower yield of $Y_c = 0.8$. The lower yield of this class of
compounds was attributed at least in part to growth inhibition by
these compounds [113, 114].

During growth on hydrocarbons the yield of cells, based on
available electrons in the substrate, $Y_{ave}-$, is lower than the
general value of 3.14. This has been attributed by Abbot and
Gledhill [114] to uncoupled oxidations in the initial alkane hydroxylation reaction. Similar uncouplings between oxidation and energy
generation exist in the initial oxidation of phenolics by mixed
function oxidases. This explanation for varying yields should help
accommodate a number of natural substrates in the generalized theory.

The applicability of cell yield functions, based either on
substrate consumed, Y_{sub}; on carbon consumed, Y_c; on reducing power
utilized, $Y_{ave}-$; or on calories utilized for growth relative to
calories in metabolites and in the initial substrate, Y_{cal}; is the
center of a fairly active discussion in the microbial biochemistry
literature [105, 112, 114]. Proof of its applicability to complex
substrates and growth other than in liquid culture would prove most
useful in studying microbial growth in nature.

C. Microbial Growth in Nature

A typical bacterial cell composition on a dry weight basis is
53% C, 7.3% H, 12% N, 19% O, and 8% ash [115]. On the basis of
Y_{ave}- = 3.14, glucose, which has 24 available electrons, should yield
75.4 g dry wt cells or 84.4 g dry wt on an ash-free basis. This
yields the figure of Y_c = 84.4/72 = 1.1 [113]. The yield based on
the carbon contents of the cell and substrate should be approximately
60%. This is considerably higher than the efficiency of growth
usually quoted for soil microorganisms. However, most of the earlier
conclusions regarding soils were drawn from work in which nonlabeled
glucose was added to soil. There are so many reactions involved in
soil that the decomposition of glucose in such systems as soil or
sediment can only be measured accurately if tracer carbon is used to
separate the degradation of added substrate from that of the naturally
occurring material.

Wagner, in Chapter 6 of this volume, applies the yield concepts
described above to calculate microbial growth of a normal soil population on ^{14}C-glucose. All carbon not respired was assumed to be
present as microbial cells. Other literature data for ^{14}C-labeled
glucose and acetate decomposition rates in soil are summarized in
Fig. 4. References and background data are presented in Table 2.
Data for very short-term experiments were not included. There are
a number of factors determining the amount of added carbon remaining
in soil after incubation; efficiency of incorporation into microbial
cells is only one of these. The curves indicate that the percent
added carbon remaining after the first flush of activity does not
appear to be related to soil texture, previous land use, or amount
of added substrate. This observation agrees with Jenkinson's [93]
conclusions when he reviewed the work on degradation of labeled
plant residues in soil.

.   In the glucose experiments shown in Fig. 4 the initial lag
period varied from 0 to 2 days, with the length of this lag having
no apparent effect on the final point of stabilization of the added
carbon. The two experiments utilizing acetate as a substrate gave

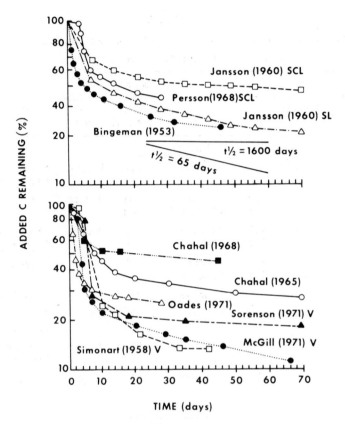

FIG. 4. Stabilization of ^{14}C-labeled glucose and acetate C added to soils in the laboratory.

among the lowest recoveries of ^{14}C remaining in the soil. As will be discussed later in this section, McGill et al. [123] did not attribute this low recovery to a low efficiency of microbial growth on this substrate. These workers attributed the low apparent efficiency to the sequential growth of a number of microbial populations, each having a fairly high efficiency. However, the sequential growth resulted in the loss of a large amount of the labeled carbon before the microbial population and its metabolites became stabilized in the soil. The amount of labeled carbon retained in the soil after the first flush of activity therefore depends on the point of stabilization of the microbial biomass and

TABLE 2

Laboratory Investigations in which ^{14}C-Labeled Glucose and
Acetate Substrates Were Added to Soil

Soil properties				Amount added (mg C/100 g soil)	Previous land use	Data from
Texture[a]	Organic C	pH	Amendment			
Mineral soils						
SL	3.0	6.5	Glucose	250	Virgin	Simonart and Mayaudon [116]
SL	1.8	6.9	Glucose	62	Virgin	Jansson [117]
SCL	2.3	5.1	Glucose	62		
SiL	0.8	4.5	Glucose	250	Wheat	Chahal and
SiL	1.4	5.0	Glucose	250	Timothy	Wagner [118]
SL	1.8	6.9	Glucose	250		Persson [119]
SCL	2.3	5.1	Glucose	250		
SL		4.8	Glucose	250		Chahal [120]
SiL	1.2	6.0	Glucose	125	Corn	Oades and Wagner [121]
C	1.7	7.6	Acetate	500	Virgin	Sorensen and Paul [122]
C	2.9	7.8	Acetate	400	Virgin	McGill et al. [123]
Organic soils						
Peat	44.5	4.9	Glucose	240	Virgin	Bingeman et al. [124]

[a]SL = Sandy Loam; SCL = Sandy Clay Loam;
SiL = Silt Loam; C = Clay.

its metabolites and is not necessarily an index of growth efficiency.

Measurements of the actual glucose or acetate remaining in soil
during incubation indicate that the original substrate disappears
rapidly [119, 125]. Under aerobic incubation conditions, such as
those existing in most terrestrial soils, few intermediate products
of metabolism, such as those occurring during the fermentation process, would be expected to accumulate. The work on microbial growth
yields [105-114], however, has not to date considered the status of

capsular materials and extracellular polysaccharides that can accumulate in a system such as soil.

In Fig. 4, respiration of labeled carbon during the first 7 days accounted for 35-55% of the added carbon in the cultivated mineral soils studied, except where acetate was added. The slopes of the lines representing half-lives in the figure indicate the great resistance to further degradation of the stabilized carbon, with the half-life of the materials remaining after 20 days ranging from 65 to 1600 days.

The relationship among microbial populations, initial substrate utilization, and later turnover of the microbial population and its metabolites has been investigated in a series of laboratory and field experiments [123, 125, 126] in which ^{14}C and ^{15}N contents of both organic and inorganic components of the system, CO_2 production, and microbial numbers were determined on the same samples. This made it possible to follow the transformations undergone by a labeled population developed on radioactive acetate and labeled $(NH_4)_2SO_4$ *in situ*. The mathematical description of the relationships existing among microbial growth, CO_2 evolution, and carbon and nitrogen turnover in the soil was based on the following concepts [123, 126].

1. Two biochemically separate populations developed sequentially.

2. The primary population was defined as that population which was the sink for added C, but that also assimilated more complex components of microbial growth.

3. The secondary population was defined as that population which utilized microbial metabolites and soil organic matter but not added C.

4. All populations were assumed to undergo some cryptic growth.

5. The quantity of C and N entering a given population was a function of the biomass C in that population.

6. The quantity of C or N released from a given population was

dependent on the amount present.

7. Organic N turnover was strictly dependent on C turnover.

8. The relative proportions of soil C and labeled C (or N) in a population could be determined by using the specific activity and quantity of evolved CO_2 as the experimental input to the model.

9. Plate count data were used to provide an index of bacterial activity with 100% activity being considered to occur when the plate count numbers reached an upper plateau. This plateau was observed during the period after day 10. The values obtained from the plate count prior to day 10 (numbers, biomass C or N) were then calculated as a percent of that value. These percentage values were then multiplied by the direct count values (total population) to obtain a quantitative estimate of the viable bacterial biomass.

Carbon dioxide evolution rates and specific activities were used as experimental inputs to the model. To check the fit of the calculations and constants used, the predicted mineral N levels and atom percent excess ^{15}N values were compared with experimental values obtained during the laboratory incubation from which the CO_2 evolution data were acquired. The model predicted mineral N and ^{15}N levels in accordance with experimental data. Therefore, it was felt that the concepts and mathematical relationships used to develop the model were essentially correct and that the model was a reasonable approximation of C and N turnover in the soil system under investigation. The distribution of labeled and nonlabeled carbon and nitrogen and their movement through the various microbial populations were calculated on the basis of the model described in Fig. 5. At time t = 0, the system contained organic soil C, labeled organic C (acetate), organic soil N, labeled mineral N, and unlabeled soil mineral N, plus a soil population. The labeled carbon was partitioned between the acetate and the microbial pool. Population dynamics were studied by measuring N mineralization and immobilization and CO_2 evolution. Microbial metabolites measured after hydrolysis with 6 N HCl or a sodium pyrophosphate, ultrasonic, water-dispersion technique indicated the relationship of various metabolites within the soil components [123, 125, 126].

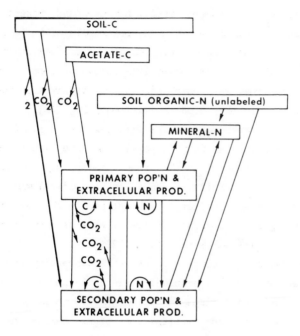

FIG. 5. Conceptual model of C and N transfer through various microbial and soil components.

Microscopic examination indicated that a predominantly fungal population developed rapidly followed by a mixed population of fungi and bacteria. Computer simulation of the above system established that the first population had a C:N ratio of 15, with a large amount of extracellular C. The metabolites from this population were attacked and converted into the second population with a C:N ratio of 5. The model gave most accurate estimates of nitrogen mineralization rates when it was assumed that both populations incorporated the added C into tissue and metabolites with an efficiency of 60%.

The estimation of microbial carbon by direct microscopy and calculation of ^{14}C biomass carbon could only account for 10% of the ^{14}C remaining in the soil, with the rest being present as uncountable dead cells, extracellular metabolites, and lytic products of the soil flora. Fractionation indicated that at the time of maximum microbial growth the acid-soluble fraction of the soil released by pyrophosphate extraction was composed of low molecular weight materials. These were rapidly attacked and converted to new microbial biomass. After sonication, components of the microbial biomass were also partitioned among various sedimentation fractions of the soil. Cell wall and particulate components, concentrated in the >0.2-μm fraction, were more resistant to attack than the cytoplasmic materials in the <0.04-μm fraction [123, 126].

In a similar experiment in the field, but with glucose as the initial substrate for the *in situ* population, the carbon loss was only 40-50% at the time of initial stabilization of the amounts of tracer. Much of the stabilized carbon and associated nutrients in the biomass or its metabolites is susceptible to chloroform treatment, indicating the importance of the biomass as a reservoir of nutrients in the soil subsystem [125, 127].

A number of other detailed experiments concerning microbial growth and activities in the soil are presently underway in forest, tundra, grassland, estuarine, and marine environments [2, 128-131]. The coordinated, detailed investigations of microorganisms and their activities in the field involve the measurement of cellulose decomposition rates, enzyme activity, and respiration, together with biomass estimates obtained by ATP or microscopic means. These should make possible a better understanding of the growth and activity of organisms in the soil subsystem.

VI. CONCLUDING REMARKS

The field of soil biochemistry has involved primarily the chemistry of soil components of biologic origin. However, an understanding of factors causing movements of nutrients, such as carbon, nitrogen, phosphorus, and sulfur, through the live soil population and the ecosystem generally requires a broader approach. In this chapter only a small portion of the nitrogen cycle and some of the challenges involved in investigating microbial growth and activities in the soil are sampled. Other examples might have served equally well to indicate the close relationships among soil biochemistry, microbial ecology, and ecosystems research.

Scientific progress is probably equally dependent on the availability of usable assay procedures and on new concepts. This is well exemplified by the surge of data in the field of nitrogen fixation and can be attributed to the development of the acetylene assay procedure for measuring nitrogenase activity [132, 133]. Microbial growth kinetics, enzyme reactions, nutrient turnover, and cell metabolite stabilization, to eventually form the more resistant soil organic matter components, are complex phenomena. No one technique or approach can both measure and describe these phenomena in nature let alone determine their significance. Enzymatic studies can be separated into two classes. One is the measurement of soil enzymes the activity of which is not necessarily related to the presence of an active energy producing system. Cellulases, proteases, and ureases in soil are examples. These enzymes act individually and can be complexed with soil humic substances and adsorbed to inorganic particles. Large background levels can build up. A second class of enzymatic reactions requires a metabolizing cell or its active constituents because energy transport is involved. Nitrogenase is an example of such a system [133]. Other examples are the measurements of activity involving electron transport in aquatic systems [134].

Further progress will doubtless be based on tracer studies and direct observation of the soil. Direct observation can be quite specific, as with use of immunofluorescence [135], or more general, as with light or scanning electron microscopes [136]. An interesting example of the quantification of visual techniques is the use of electron probe microanalyses to investigate the cation composition of decomposer organisms *in situ* [137]. Derivation of mathematical models and use of computers for data handling will help describe the operation of any system under study.

REFERENCES

1. E. P. Odum, *Fundamentals of Ecology*, 3rd ed., p. 8, Saunders, Philadelphia, 1971.

2. M. Witkamp, *Ann. Rev. Ecol. Syst.*, *2*, 85 (1971).
3. D. E. Reichle, *Analyses of Temperate Forest Ecosystems*, Springer-Verlag, Berlin, 1970.
4. G. Van Dyne, *The Ecosystem Concept in Natural Resource Management*, Academic, New York, 1969.
5. J. D. Strickland, in *21st Symp. Soc. Gen. Microbiol.*, p. 280, University Press, Cambridge, 1971.
6. G. Sykes and F. A. Skinner, *Microbial Aspects of Pollution*, Academic, New York, 1971.
7. N. A. Krasil'nikov, *Soil Microorganisms and Higher Plants* (Eng. Transl.), Academy of Sciences, USSR, Moscow, 1958.
8. C. D. Bowen and A. D. Rovira, in *Root Growth* (W. J. Whittington, ed.), p. 170, Butterworths, London, 1969.
9. H. Jenny and K. Grossenbacher, *Soil Sci. Soc. Amer. Proc.*, *27*, 273 (1963).
10. J. F. Darbyshire and M. P. Greaves, *Soil Biol. Biochem.*, *3*, 151 (1971).
11. G. D. Bowen and A. D. Rovira, *Nature (London)*, *211*, 665 (1966).
12. R. H. Miller and T. J. Chau, *Plant Soil*, *32*, 146 (1970).
13. D. A. Barber and B. C. Loughman, *J. Exp. Bot.*, *18*, 170 (1967).
14. D. A. Barber, J. Sanderson, and R. Russell, *Nature (London)*, *217*, 644 (1962).
15. O. E. Bradfute, R. A. Luse, L. Braal, and A. D. McLaren, *Soil Sci. Soc. Amer. Proc.*, *26*, 406 (1962).
16. M. E. Brown, *J. Appl. Bact.*, *35*, 443 (1972).
17. A. Pvapst and M. Schnitzer, *Soil Biol. Biochem.*, *3*, 215 (1971).
18. W. Flaig, in *Symposium on Use of Isotopes in Agriculture and Animal Husbandry Research*, IAEA, New Delhi, 1971.
19. J. L. Neal, Jr., *Can. J. Microbiol.*, *15*, 633 (1969).
20. F. E. Clark and E. A. Paul, *Advan. Agron.*, *22*, 375 (1970).
21. E. L. Rice and S. K. Parncholy, *Amer. J. Bot.*, *59*, 1033 (1972).
22. S. S. Brar and J. Giddens, *Soil Sci. Soc. Amer. Proc.*, *32*, 821 (1968).
23. J. B. Robinson, *Plant Soil*, *19*, 173 (1963).
24. J. L. M. Huntjens, *Plant Soil*, *34*, 393 (1971).
25. J. L. M. Huntjens, *Plant Soil*, *35*, 77 (1971).
26. W. V. Bartholomew and F. E. Clark, *Soil Nitrogen*, pp. 314-318, Amer. Soc. Agron. Ser. 10, Madison, 1965.
27. W. Verstraete and M. Alexander, *J. Bact.*, *110*, 955 (1972).
28. M. Alexander, *Ann. Rev. Microbiol.*, *25*, 361 (1971).
29. G. H. Elkan, *Can. J. Microbiol.*, *8*, 79 (1962).
30. L. R. Mytton and D. Gareth Jones, in *Biological Nitrogen Fixation in Natural and Agricultural Habitats* (T. A. Lie and E. G. Mulder, Eds.), p. 17, Martinus Nijhoff, The Hague, 1971.
31. J. L. Neal, Jr., T. G. Atkinson and R. I. Larson, *Can. J. Microbiol.*, *16*, 153 (1970).
32. Y. Abd-el-Malek, in *Biological Nitrogen Fixation in Natural and Agricultural Habitats* (T. A. Lie and E. G. Mulder, Eds.), p. 425, Martinus Nijhoff, The Hague, 1971.
33. J. Dobereiner and A. B. Campelo, in *Biological Nitrogen Fixation in Natural and Agricultural Habitats* (T. A. Lie and

E. G. Mulder, Eds.), p. 457, Martinus Hijhoff, The Hague, 1971.

34. D. L. Kass, M. Drosdoff, and M. Alexander, *Proc. Soil Sci. Soc. Amer.*, *35*, 286 (1971).

35. J. Dobereiner, J. M. Day, and P. J. Dart, *J. Gen. Microbiol.*, *71*, 103 (1972).

36. H. J. Evans, N. E. R. Campbell, and S. Hill, *Can. J. Microbiol.*, *18*, 13 (1972).

37. J. Macura, *Folia Microbicl.*, *16*, 328 (1971).

38. D. Parkinson, in *Soil Biology* (N. A. Burgess and F. Raw, Eds.), p. 449, Academic, New York, 1967.

39. W. Gams, *Mitt. Biol. Bundesans. Land.-Fortswirt.*, Berlin, *123*, 1 (1967).

40. H. Katznelson and J. W. Rouatt, *Can. J. Microbiol.*, *3*, 673 (1957).

41. N. J. Brown, E. R. Fountaine, and M. R. Holden, *J. Agron. Sci.*, *64*, 195 (1965).

42. C. L. Kucera and D. R. Kirkham, *Ecology*, *52*, 912 (1971).

43. F. Warembourg and E. A. Paul, *Plant Soil* , *38*, 331 (1973).

44. H. J. V. Schappert, M.Sc. Thesis, University of Saskatchewan, Saskatoon, 1972.

45. A. D. McLaren, *Science*, *141*, 114 (1963).

46. T. R. G. Gray and S. T. Williams, *Soil Microorganisms*, Oliver and Boyd, Edinburgh, 1971.

47. S. T. Williams and C. I. Mayfield, *Soil Biol. Biochem.*, *3*, 197 (1971).

48. C. I. Mayfield, S. T. Williams, S. M. Ruddick, and H. L. Hatfield, *Soil Biol. Biochem.*, *4*, 79 (1972).

49. A. D. McLaren and L. Packer, *Advan. Enzymol.*, *33*, 245 (1970).

50. M. Nishio and C. Furusaka, *Soil Sci. Plant Nutr.*, *17*, 54, 61 (1971).

51. M. Naito, T. Takamatsu, L. T. Fau, and E. S. Lee, *Biotech. Bioeng.*, *11*, 731 (1969).

52. E. A. Falch and E. L. Gaden, *Biotech. Bioeng.*, *12*, 465 (1970).

53. A. Prokop, L. E. Erickson, J. Fernandez, and A. E. Humphrey, *Biotech. Bioeng.*, *11*, 945 (1969).

54. A. D. McLaren, *Soil Sci. Soc. Amer. Proc.*, *33*, 551 (1969).

55. C. M. Cho, *Can. J. Soil Sci.*, *51*, 339 (1971).

56. H. C. Preul and G. J. Schroepfer, *J. Water Poll. Control Fed.*, *40*, 30 (1968).

57. S. Ardakani, J. Rehbock and A. D. McLaren, *Soil Sci. Soc. Amer. Proc.*, *37*, in press (1973).

58. K. Sato, *Bull. Inst. Agri. Res. Tohoku Univ.*, *22*, 93 (1971).

59. C. C. Delwiche and P. L. Steyn, *Environ. Sci. Technol.*, *4*, 929 (1970).

60. D. H. Kohl, G. Shearer, and B. Commoner, *Science*, *174*, 1331 (1971).

61. D. H. Cheng, J. M. Bremner, and A. P. Edwards, *Science*, *146*, 1574 (1964).

62. J. M. Bremner and M. A. Tabatabai, *J. Env. Qual.*, *2*, 363 (1973).

63. M. H. Briggs and L. Segal, *Life Sci. (Oxford)*, *69* (1963).

64. M. Martin-Smith, *Nature (London)*, *197*, 367 (1963).

65. R. Bartha and L. Bordeleau, *Soil Biol. Biochem.*, *1*, 139 (1969).
66. L. W. Getzin and I. Rosefield, *Biochim. Biophys. Acta*, *235*, 442 (1971).
67. M. A. Chalvignac and J. Mayaudon, *Plant Soil*, *34*, 25 (1971).
68. R. G. Burns, M. H. El-Sayed, and A. D. McLaren, *Soil Biol. Biochem.*, *4*, 107 (1972).
69. I. Tanabe and S. Ishizawa, *Bull. Natl. Inst. Agron. Sci. (Japan)*, Series B, No. 21 (1969).
70. A. D. McLaren and J. Skujins, *The Ecology of Soil Bacteria* (T. R. Gray and D. Parkinson, Eds.), p. 15, Univ. of Toronto Press, Toronto, 1968.
71. E. F. Estermann, G. H. Peterson, and A. D. McLaren, *Proc. Soil Sci. Soc. Amer.*, *23*, 31 (1959).
72. J. N. Ladd and J. H. A. Butler, *Soil Biol. Biochem.*, *3*, 157 (1971).
73. A. D. McLaren, *Compt. Rend. Laboratoire Carlsberg*, *28*, 175 (1952).
74. D. Murphy and A. W. Moore, *Sci. Proc. Roy. Dublin Soc.*, *1-A*, 191 (1960).
75. A. Burges, *Sci. Proc. Roy. Dublin Soc.*, *1-A*, 53 (1960).
76. J. Bremner, in *Soil Biochemistry* (A. D. McLaren and G. H. Peterson, Eds.), Vol. 1, p. 19, Dekker, New York, 1967.
77. P. Simonart, L. Batistic, and J. Mayaudon, *Plant Soil*, *27*, 153 (1967).
78. V. O. Biederbeck and E. A. Paul, *Soil Sci. 115*, 357 (1973).
79. M. A. Tabatabai and J. M. Bremner, *Soil Biol. Biochem.*, *3*, 317 (1971).
80. R. G. Wiegert, D. C. Coleman, and E. P. Odum, in *Methods of Study in Soil Ecology*, p. 93, IBP-UNESCO, 1971.
81. F. B. Golley, *Ecology*, *42*, 581 (1961).
82. K. W. Cummins and J. L. Wuycheck, *Int. Assoc. Theoret. Appl. Limnol., Communications*, *18*, 1 (1971).
83. G. J. Prochazka, W. J. Payne, and W. R. Mayberry, *J. Bact.*, *104*, 646 (1970).
84. E. Gorham and J. Sanger, *Ecology*, *48*, 492 (1967).
85. Y. Martel, Ph.D. Thesis, University of Saskatchewan, Saskatoon, 1972.
86. C. Kunze, *Zentr. Bakt. Pariset Infect. Hyg.*, *125*, 385 (1971).
87. O. Holm-Hansen and C. R. Booth, *Limnol. Oceanog.*, *11*, 510 (1966).
88. M. Witkamp, in *Modern Methods in the Study of Microbial Ecology* (T. Rosswall, Ed.), p. 179, Bull. Ecol. Res. Comm., Swedish National Res. Council (Stockholm), Vol. 17, 1973.
89. B. S. Ausmus, in *Modern Methods in the Study of Microbial Ecology* (T. Rosswall, Ed.), Vol. 17, p. 223, Bull. Ecol. Res. Comm., Swedish National Res. Council (Stockholm), 1973.
90. L. C. Lee, R. F. Harris, J. D. H. Williams, D. E. Armstrong, and J. K. Syers, *Soil Sci. Soc. Amer. Proc.*, *35*, 86 (1971).
91. G. M. Woodwell and D. B. Botkin, in *Analysis of Temperate Forest Ecosystems* (D. E. Reichle, Ed.), p. 73, Springer-Verlag, 1970.

92. E. de Jong and H. J. V. Schappert, *Soil Sci.*, *113*, 328 (1972).
93. D. S. Jenkinson, *Soil Sci.*, *111*, 64 (1971).
94. R. C. Dahlman and C. L. Kucera, *Ecology*, *49*, 1199 (1968).
95. E. A. Paul and F. R. Warembourg, in *Modern Methods in the Study of Microbial Ecology* (T. Rosswall, Ed.), Vol. 17, p. 274, Bull. Ecol. Res. Comm., Swedish National Res. Council (Stockholm), 1973.
96. J. E. Hobbie and C. C. Crawford, *Limnol. Oceanog.*, *14*, 528 (1969).
97. K. J. Hall, P. M. Kleiber, and I. Yesaki, *IBP-UNESCO Symposium on Detritus and Its Ecological Role in Aquatic Systems*, (V. Melchiorri-Santolini and J. Hopton, Eds.) Memorie Dell' Instituto Italiano di Idrobiologia, Suppl., *29*, 1972.
98. IAEA, *Nitrogen in Soil-Plant Studies*, Vienna, 1971.
99. H. Freytag and H. Igel, *Albr. Thaer-Archiv.*, *14*, 859 (1970).
100. D. Coleman, *Mycologia*, *LX*, 960 (1968).
101. C. A. Edwards, D. E. Reichle, and D. A. Crossley, in *Analysis of Temperate Forest Ecosystems* (D. E. Reichle, Ed.), p. 147, Springer-Verlag, 1970.
102. N. Van Uden, *Ann. Rev. Microbiol.*, *23*, 473 (1969).
103. S. S. Pirt and W. M. Kurowski, *J. Gen. Microbiol.*, *63*, 357 (1970).
104. K. M. Peil and A. F. Gaudy, *J. Appl. Microbiol.*, *21*, 253 (1971).
105. W. W. Forrest and D. J. Walker, *Advan. Microbiol. Phys.*, *5*, 213 (1971).
106. E. Umbarger and B. D. Davis, in *The Bacteria* (I. C. Gunsalus and R. Y. Stanier, Eds.), Vol. 3, p. 167, Academic, New York, 1962.
107. J. Mandelstam and K. McQuillen, *Biochemistry of Bacterial Growth,* Wiley, New York, 1968.
108. B. Magasanik, in *The Bacteria* (I. C. Gunsalus and R. Y. Stanier, Eds.), Vol. 3, p. 295, Academic, New York, 1962.
109. H. J. Morowitz, *Energy Flow in Biology*, Academic, New York, 1968.
110. T. Bauchop and S. R. Elsden, *J. Gen. Microbiol.*, *23*, 457 (1970).
111. W. R. Mayberry, G. J. Prochazka, and W. J. Payne, *Appl. Microbiol.*, *15*, 1332 (1967).
112. W. J. Payne, *Ann. Rev. Microbiol.*, *24*, 17 (1970).
113. G. H. Bell, *Process Biochem.*, *7*, 21 (1972).
114. B. J. Abbot and W. E. Gledhill, *Advan. Appl. Microbiol.*, *14*, 249 (1971).
115. A. E. Humphrey, in *Single Cell Protein* (R. E. Mateles and S. R. Tannenbaum, Eds.), p. 330, MIT Press, Cambridge, 1968.
116. P. Simonart and J. Mayaudon, *Plant and Soil*, *9*, 367 (1958).
117. S. L. Jansson, *Trans. 7th Int. Cong. Soil Sci.*, *2*, 635 (1960).
118. K. S. Chahal and G. H. Wagner, *Soil Sci.*, *100*, 96 (1965).
119. J. Persson, *LantbrHögsk. Ann.*, *3*, 81 (1968).

120. K. S. Chahal, in *Isotopes and Radiation in Soil Organic Matter Studies*, p. 207, IAEA, Vienna, 1968.
121. J. M. Oades and G. H. Wagner, *Soil Sci. Soc. Amer. Proc.*, *35*, 914 (1971).
122. L. H. Sorensen and E. A. Paul, *Soil Biol. Biochem.*, *3*, 173 (1971).
123. W. B. McGill, E. A. Paul, J. A. Shields, and W. E. Lowe, in *Modern Methods in the Study of Microbial Ecology* (T. Rosswall, Ed.), Vol. 17, p. 293, Bull. Ecol. Comm., Swedish National Res. Council (Stockholm), 1973.
124. C. W. Bingeman, J. E. Varner, and W. P. Martin, *Soil Sci. Soc. Amer. Proc.*, *17*, 34 (1953).
125. J. A. Shields, E. A. Paul, W. E. Lowe, and D. A. Parkinson, *Soil Biol. Biochem.*, *5*, 753 (1973).
126. W. B. McGill, Ph.D. Thesis, University of Saskatchewan, Saskatoon, 1972.
127. D. S. Jenkinson, *J. Soil Sci.*, *17*, 280 (1966).
128. D. Parkinson, *Progress in Soil Microbiology*, Report, IBP-UNESCO Meeting, Paris, 1970.
129. F. Bunnell, in *Modern Methods in the Study of Microbial Ecology* (T. Rosswall, Ed.), Vol. 17, p. 407, Bull. Ecol. Res. Comm., Swedish National Res. Council (Stockholm), 1973.
130. T. V. Aristovskaya, *Problems of Abundance, Biomass and Productivity of Microorganisms in Soil* (in Russian with English summaries), Nauka, Leningrad, 1972.
131. IBP-UNESCO, *Symposium on Detritus and Its Ecological Role in Aquatic Ecosystems*, (U. Melchiorri-Santolini and J. Hopton, Eds.), Memorie Dell' Istuto Italiano di Idrobiologia, Suppl., *29*, 1972.
132. R. W. F. Hardy, R. C. Burns, and R. D. Holsten, *Soil Biol. Biochem.*, *5*, 47, 1973.
133. H. Dalton and L. E. Mortensen, *Bact. Rev.*, *36*, 231 (1972).
134. T. Packard and M. L. Healy, *Limnol. Oceanog.*, *16*, 60 (1968).
135. E. L. Schmidt, in *Modern Methods in the Study of Microbial Ecology* (T. Rosswall, Ed.), Vol. 17, p. 67, Bull. Ecol. Res. Comm. (Stockholm), 1973.
136. B. Berg, B. Hofsten, and G. Petersen, *J. Appl. Bact.*, *35*, 215 (1972).
137. R. L. Todd, K. Cromack, and J. C. Stormer, Jr., *Nature (London)*, *243*, 544 (1973).

CHAPTER 2

SYSTEMS ANALYSIS OF SOIL ECOSYSTEMS

J. S. Russell

CSIRO
Division of Tropical Agronomy, Cunningham Laboratory
St. Lucia, Queensland, Australia

I. INTRODUCTION

The soil is an important component of most terrestrial ecosystems and of the water and biogeochemical cycles and the energy
flow cascade. The soil also provides mechanical support for vegetation and is a habitat for macro- and microorganisms. The interactions of the climate-soil-plant-animal complex are such that it is
difficult to integrate various components of even relatively simple

systems. The development of the computer, with its capacity for
numerical manipulation, is providing opportunities for the study of
complex systems. In particular, systems analysis is an approach
that is of value in understanding ecosystems and in gaining insight
into their behavior.

A. Systems Analysis

"Systems analysis" and "ecosystems" are terms that have become
widely used in recent years. Initially, "systems analysis" was
considered to mean an analysis of the interrelationships among
significant components of a system. The boundaries of the system
were defined by the analyst and there could be various subsystems
within an overall system. However, recent usage of the term in some
fields of science has given a broader meaning than this. It now
involves showing relationships among components in mathematical
terms in the form of a model and it is this aspect of systems
analysis that is considered in this chapter.

After the mathematical model has been constructed it is programmed for a computer and operated to simulate the system. The
model output is then compared with data of the real system. Model
validation may involve a number of recursive steps in which adjustments are made to the model either conceptually or mathematically
until an adequate model is obtained. The various steps involved in
this type of systems analysis are shown in Fig. 1.

The systems analysis approach can be compared to that of
problem solving. Ross [1] distinguished four phases in man-machine
problem solving. These are (a) the lexical phase, in which the
problem is broken into discrete items; (b) the parsing phase, in
which the items are grouped into a structure; (c) the modeling
phase, in which meaning is extracted from the structure and an
understanding of the problem formulated; and (d) the analysis phase,
in which the problem is solved. Dale [2] suggests that these four
phases can also be identified in systems analysis as follows: (a)
determination of the entities or parts, (b) the choice of

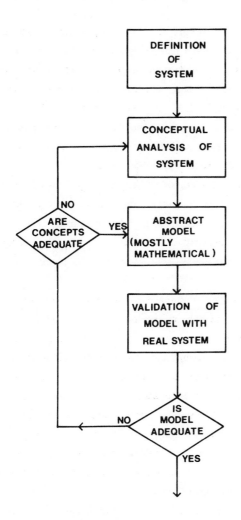

FIG. 1. Sequential diagram of the systems analysis approach.

relationships between those that are of interest, (c) the specifica-
tion of the mechanism by which these interrelationships take place,
and (d) the validation of the model of the system so produced and
investigation of its properties.

In modeling, an attempt is made to simulate the real system
either to increase understanding or to gain control over some of
the variables. Correspondence over a wide range of variables is

usually desirable but not essential. So long as the model simulates
the real system over a definite range and outputs of the model and
the real system are highly correlated, then the model may be useful
even though the mechanisms of the model do not reflect those of the
real system. Because most models involve subsystems it is desirable
that outputs from the subsystems be also highly correlated with real
subsystem outputs. The most satisfactory form of model validation
is one that involves validation of subsystems.

Because of its emphasis on model building and model validation,
systems analysis is directly dependent on the computer. The first
three steps of Fig. 1 are no different from research practice in the
past and can be readily carried out in the absence of a computer.
However, for other than a very simple system, the remaining step
cannot.

With the development of computer technology there has been a
gradual increase in the breadth of analysis (from the specialized to
the general) and the depth of analysis (from the simple to the complex). With the early technology it was only possible to analyze
simple general systems or complex specialized systems. Later developments have assisted the ultimate aim, that is, the analysis of
complex general systems. There has been a gradual evolution in
digital computers from assembly languages to machine languages to
simple languages, e.g., FORTRAN (Formula Translation) to specific
modeling languages, e.g., CSMP (Continuous Systems Modeling Programs) to interactive systems [3, 4].

The development of specific modeling languages, such as CSMP
[5], has simplified the programming required for mathematical models
involving differential equations. The utility of such a language
in the analysis of complex natural systems has been shown by Brennan
et al. [6]. However, FORTRAN can be used for comparatively simple
systems [7]. In addition subroutines in FORTRAN are available for
continuous systems modeling, e.g., [8] and these do not have the
stringent requirements associated with specific modeling languages,
which may be restricted to certain large computers.

Some of the early systems analysis studies were carried out with analog computers. These have some advantages, particularly in relation to conceptual operation and interactive properties, but the large digital computer is more versatile and has shown more potential for evolutionary change. Of the eight models to be discussed in some detail in this chapter, five were programmed for digital computers and three for analog computers. Of the five studies on the digital computer, four were programmed in FORTRAN and one in CSMP.

The quantitative approach of systems analysis emphasizes the need both for data and for the mathematical expression of relationships. Emshoff and Sisson [9] have pointed out that, in simulation, data are needed for the estimation of values of constants and parameters, for starting values for all variables, and for validating the model.

The mathematical nature of systems analysis models may vary widely. Clymer [10] has suggested that models may be classified in terms of where they are located along a number of axes comprising model space. Some of these axes, such as continuous-discrete or deterministic-stochastic, specify the mathematical approach that has been used in describing the system. Milsum [11] has distinguished three broad classes of mathematical representation in models. These are: (1) sets of algebraic equations; (2) sets of ordinary differential equations; and (3) sets of partial differential equations.

Discrete models generally consist of algebraic equations only, whereas continuous models usually involve differential equations. Some of the numerical methods for the solution of sets of ordinary differential equations have been reviewed by Benyon [12]. In practice, sets of as many as 700 differential equations have been solved numerically in physical systems. Nothing of this complexity has yet been developed for biologic systems.

B. Soil Ecosystems

1. Definition

The concept of the ecosystem was originally introduced by

Tansley [13] as a holistic unit to include not only the vegetation but also the environment of the vegetation, including climate and soil. Therefore, the ecosystem includes not only the organism complex but the whole array of physical factors forming the environment of the biome. Ecosystems show organization that is the result of interaction of their components and there is a tendency for the attainment of a dynamic equilibrium in ecosystems between the inorganic factors and the organisms.

Dale [2] has defined an ecosystem as a system open for at least one property and in which at least one of the entities is classed as living. Thus, soil ecosystems are those systems in which there is an interaction between soil physical and chemical factors and organisms, including microorganisms, vegetation, and man.

Soil ecosystems have a number of distinctive characteristics that affect their analysis. These are

a. The complexity of soil is such that even basic knowledge of the components of the ecosystem and their interrelationships is not available. This is common to all biologic systems but it is intensified in soils by the difficulties of analysis and measurement. To study soil-plant interactions, for example, it is frequently necessary to destructively sample or disturb the soil, which makes information on rates of processes particularly difficult to obtain.

b. Soil is a continuum and many soil entities can be defined. Knox [14] has pointed out that soil entities may range from primary particles of sand, silt, and clay to landscape units. Although some of the entities are isotropic, others, such as soil profiles and pedons, are anisotropic and multidimensional.

c. Layering of soil presents difficulties in the mathematical representation of interrelationships. This problem also exists in the numerical taxonomic analysis of soils and various methods can be used to represent the vertical change in soil properties [15, 16]. In systems analysis of soils a number of approaches have been used. These are shown in Fig. 2 and vary from considering the soil as a

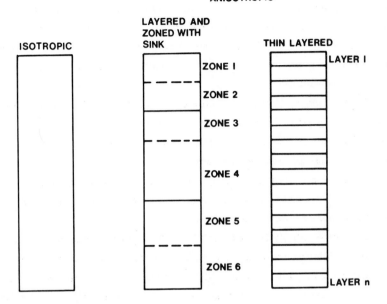

FIG. 2. Three representations of soil profiles with increasing requirement for parameter definition.

simple isotropic system to considering it as a many-layered aniso-tropic system. The difficulties of computation and parameter definition increase with increasing delineation of layering. For some ecosystems consideration of the soil as isotropic may be a justifiable assumption. However the level of modeling being at-tempted and the possibility of depth interactions are important in deciding the method of representation of the soil profile.

2. Grouping of Ecosystems

Because of the holistic nature of the ecosystem concept any grouping must be artificial to some extent. In the growth of plants, water, nutrients, and energy are all required and in a detailed model all would have to be represented. Nevertheless, it is possible to recognize situations in which certain components are limiting and others are nonlimiting. For instance, in most semiarid environments water is the main limiting factor for plant growth. A model that adequately simulates the availability of soil water may explain a considerable proportion of the variation in plant growth.

Soil ecosystems will be considered under three broad groupings— the water or hydrologic cycle, the biogeochemical cycle, and the energy flow cascade.

This grouping considers soil ecosystems in relation to the basic driving functions of water, nutrients, and energy. In this chapter no attempt will be made to consider the population and community aspects of soil ecosystems.

Soil is an important component of all three groupings. Flow diagrams of the relationships among soil, living entities, and driving functions are shown in Fig. 3.

In the water cycle, the soil is an important component in the transfer of water from the atmosphere to the biologic components of the system, and in the biogeochemical cycle it is a major source of nutrient elements. By contrast, soil components comprise the last stages in the cascade of solar energy flow through green plants and its gradual dissipation to organic matter in the soil.

If Dale's definition of ecosystems is accepted, that is, that at least one component must be living, systems analysis of soil eco- systems can be considered as restricted to the interactions of soil physical and chemical factors and living entities. Systems analysis has been applied to various soil processes that involve physical and chemical factors. De Wit and van Keulen [17] have simulated the transport processes of water, heat, and salts in soils. The heat transfer of soils has also been simulated [18], as has the movement of chloride in soils [19]. These systems, as modeled, have no in- teraction with living components and are not considered in detail in this chapter. However, such systems could become subsystems of broader ecosystem studies by providing physical or chemical inputs.

A number of specific models will be discussed in this chapter. Some of these models (particularly the nitrogen models) overlap two of the above groupings and classification has been based on the aspect considered most relevant.

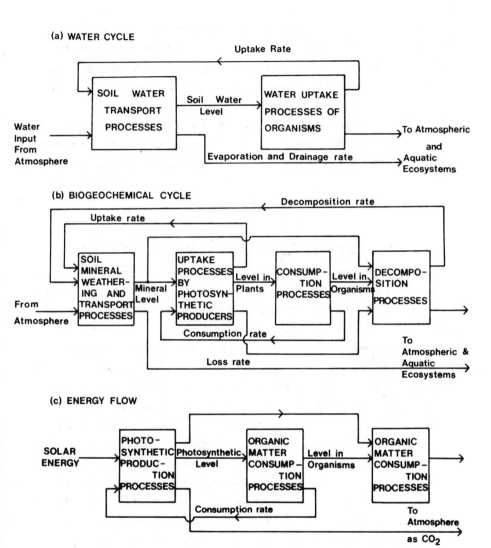

FIG. 3. Information flow diagram of some of the main processes in soil ecosystems of (a) the water cycle, (b) the biogeochemical cycle, and (c) energy flow. Boxes represent processes; lines represent variables (quantities or rates).

Two types of block diagrams of the specific models discussed in detail are presented. These are the information flow diagram [11] and the compartment diagram. Both are simultaneous, not sequential,

diagrams and are analogous to electrical circuit diagrams. Diagrams
of specific models have been redrawn to a common format and choice
of either an information flow or a compartment diagram was based
largely on the approach used in the original paper. In some cases
only the main flows have been shown.

In the information flow diagrams the boxes are processes and
the lines are quantities or rates. In the compartment diagrams the
boxes are quantities and the lines are processes or equations.

II. THE WATER CYCLE

A. Modeling of the Water Cycle and Soil Ecosystems

Systems analysis of the water cycle in soil ecosystems is at-
tractive from a number of points of view. First, there is a comprehensive theoretical and practical framework on which to base such an
analysis. Second, soil moisture fluxes are rapid and techniques of
measuring these are readily available. This facilitates model
validation. Third, there are large amounts of meteorological data of
both a current and a historical nature available for many global
stations, which could be used in adequately validated models.
Fourth, the soil water changes have a fundamental effect on both
natural and agricultural systems. If these systems are to be real-
istically simulated in the field it is necessary to have soil water
as a subsystem. Soil water and temperature largely determine the
level of biologic activity in terrestrial ecosystems. Therefore, a
model of the complex interrelationships among meteorological factors,
soil, and biologic activity would be useful in understanding these
ecosystems.

There is an extensive literature on the role of soil moisture
in the interrelationships between meteorological factors and plant
growth [20]. Computers were not used in early attempts to compile
water balances from standard meteorological observations. Much of
this work was related to the scheduling of irrigation and mostly

involved simple calculations. Less progress was made in the
estimation of soil moisture under dry land conditions. The necessity for simplifying assumptions concerning plant attributes, such
as rooting characteristics, stage of plant development, and water
requirements, and soil properties, such as storage capacity and
availability of water and the fact that some of these characteristics were time varying, posed considerable computational difficulties. With the availability of computers various soil water balance
models have evolved. Some are concerned with watershed aspects of
the hydrologic cycle, such as runoff [21-24]. Several recent models,
however, have emphasized the interaction between plant growth and
the soil and have included soil parameters to generalize the applicability of the models to a range of soils [25-30].

B. Specific Models

Two water balance models will be discussed in some detail. The
first is a discrete model of algebraic equations using a simple isotropic soil with few parameters and a simplified plant growth relationship. The second is a continuous model with a first-order
differential equation using a layered anisotropic soil with additional soil and plant parameters, the latter time varying.

1. Discrete Model

The basic outline of the Fitzpatrick et al. [25] model is shown
in an information flow diagram in Fig. 4. This model was programmed
in FORTRAN.

Three main processes were considered: runoff, soil water recharge, and soil water loss. The main inputs to the model were
rainfall and evaporation rate. Stage of growth of the vegetation
was calculated according to set criteria. In the initial model the
time step was 5 days but later versions of the model used a time
step of 7 days [28].

A two-stage soil model was used. The water stored in the root
zone at water potentials higher than -15.0 bars and regarded as

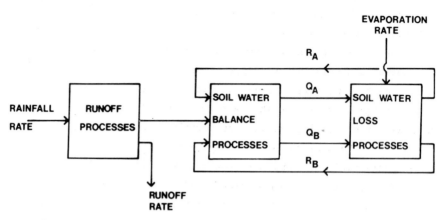

FIG. 4.   Information flow diagram of water balance model of Fitzpatrick et al [25]. Q_A and Q_B are the quantities in store A and B; R_A and R_B are the rates of water loss from store A and B.

being available for active plant growth was referred to as store A and that retained at lower potentials but above the minimum observed was called store B. The maximum amount of water in store A and store B and the proportion of rainfall added to each store during recharge were the main soil parameters. These will vary with inherent soil properties, with the depth of soil considered, and with the vegetation. In the case of communities dominated by *Eragrostis eriopada* and *Acacia aneura*, a rooting depth of 152 cm (5 ft) was used. The soil had a maximum water storage of 0.075 cm water per centimeter soil (0.9 in./ft) in store A and 0.025 cm/cm in store B, giving a total maximum storages of 11.4 and 3.8 cm of water, respectively.

The runoff function was limited by the use of daily rainfall and the time step of 5 days and simply sought to account for large amounts of rainfall over short periods. The water recharge function in the model apportioned different amounts of the rainfall to store A and B depending on certain conditions. Thus, from an initially dry condition, recharge was apportioned 75% to store A and 25% to store B, but with definite procedures at intermediate moisture contents. Soil water extraction was related to accumulated potential evaporation from the initiation of growth response and the amount of water in the soil store.

Operation of the model for periods of 50-100 years is not expensive of computing time. One of the main driving forces, daily rainfall, is available for large numbers of locations. Where daily evaporation data are not available estimates of evaporation rate can often be made from other climatic parameters [31].

2. Continuous Model

The soil model and the plant-soil interaction was very simplified in the case of the discrete model discussed in Section II,B,1. A more detailed analysis involving additional soil and plant parameters is necessary where shorter time periods and changes in the soil profile need to be considered or where inputs are required for other models. The water balance model of Ross [29] is a continuous model using soil, plant, and climatic parameters (Fig. 5). This model was programmed in FORTRAN.

In contrast to the isotropic soil model of Fitzpatrick et al., the Ross model divided the soil profile into a maximum of eight zones, which might be of variable size. The last zone was of large size and was used as a sink for water that had drained through higher zones. Up to three soil layers could be specified, each containing a number of specified zones. Soil parameter values were taken as constant with zones nested within any one layer but could differ from layer to layer. The main input variables, soil and plant parameters, used in the model are shown in Table 1.

In addition, two parameters of the roots, A and B, defined the distribution of root density with depth ($g/m^2/cm$ or $cm/m^2/cm$), with an equation of the form $A[\exp(-X/B)]$, where X is the depth in centimeters. Therefore, A was the density at the surface and B the depth at which the density had fallen to 37% of its surface value (at 3B it had fallen to 5%).

The time derivate of the volumetric moisture content of zone k, V_k, was given by

$$dV_k/dt = (R_k - R_{k+1} - A_k)/S_k \qquad (1)$$

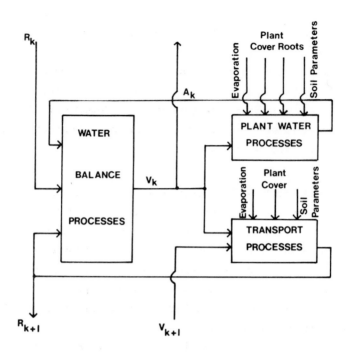

FIG. 5. Information flow diagram of the water balance model of Ross [29] for zone k. R_k is the rate of water flow between zones, A_k is the uptake rate by plants, and V_k is the volumetric water content of zone k.

where R_k for the surface zone is taken as the rainfall rate less actual evaporation from bare soil, R_{k+1} is the rate of flow from zone k to zone k + 1, A_k is the actual uptake rate by plants, and S_k is the size of zone k in centimeters. The time step of the model was 1 day.

No attempt is made in this model to account for runoff or run on. Computing time for equivalent time periods is greater than for the discrete model. Also the necessity for time-varying plant data limits the practical use of the model to shorter periods, i.e., from several months to about 5 years.

C. Aspects of Soil Water Balance Models

These two models show some of the inherent difficulties in modeling the soil-plant-water system. If a model is to be general

TABLE 1

Main Inputs to the Continuous Water Balance Model of Ross [29]

Soil parameters for each layer	Volumetric moisture content at at Matric potential field and wilting Hydraulic conductivity capacity point
Plant parameters (functions of time)	Fraction of ground covered by vegetation Fraction of vegetation green Dry weight or length of roots in each zone
Input variables (functions of time)	Rainfall rate (cm/day) Tank evaporation rate (cm/day)
Constants	Ratio of potential evapotranspiration rate from green plant parts to tank evaporation Adjustable constant for water uptake by roots Adjustable constant for rate of water movement in the soil
Initial values	Initial values of volumetric moisture content of each zone

and used on a geographic basis it must use inputs that are widely available and that use as few soil and plant parameters as possible. The discrete model fulfills these requirements. However more detailed modeling involves the use of additional soil and plant parameters, especially time-varying parameters.

For extension of water balance models to other ecosystems it is necessary for soil water processes, such as recharge, runoff, interception, and loss, to be related to specific attributes of the environment [32]. Not only should these attributes include climatic characteristics, such as rainfall and evaporation, and soil characteristics, such as texture, structure, and water conductivity, but also other characteristics, such as landform and slope and the structure and growth pattern of individuals and communities of plants. McCown [33] has shown the need for more detailed studies to define the field moisture characteristics of soils. In particular, better estimates of the upper limit of available water in poorly drained soils and of the depth of water penetration in arid soils are necessary. Furthermore, such measurements should be widely available and compatible with conventional survey methods.

Because of the significance of soil water in biologic activity and the geographic variation in climatic factors and soil properties, water balance models have a number of uses. These may be summarized as follows.

1. Simulation of Short-Term Soil Water Changes

The simulation of soil moisture with depth and time may be required in some situations. The Ross model may be of value here. Preliminary testing of this model was carried out with soil moisture data from a grey clay soil under alfalfa (*Medicago sativa*) at Meandarra (lat., 27° 20's; long., 149° 53' E; annual rainfall, 500 mm; annual evaporation, 1700 mm). Simulated values are compared with measured values over a 400-day period from October 20, 1968 (Fig. 6). The measured soil values are means of 16 profiles sampled along a transect of approximately 1 km.

Comparison of the simulated and actual values raises several problems. (1) To validate the model in detail sampling needs to be more intensive near the surface than at depth. (2) Even though the area is almost level and runoff is small the lack of a runoff function, or a means of estimating runoff, is a limitation that may lead to high simulated moisture levels in the lower horizons. (3) The rainfall for this area was measured some 3 or 4 km from the experimental site and this may also account for some of the positive or negative discrepancies in the lower soil horizons. (4) The effects of soil cracking (which is severe on these grey clays) on field soil moisture are not known and may have contributed to differences between simulated and actual values.

Undoubtedly the quality of model output depends on the quality of the climate, soil, and plant input data. Plant data in particular present difficulties. Estimation of root parameters for perennial vegetation is less difficult than for annual plants, in which rapid changes in amount and distribution occur. However, given adequate validation the use of a model such as this either as a subsystem to other models or to predict soil moisture changes should be valuable in studying soil ecosystems.

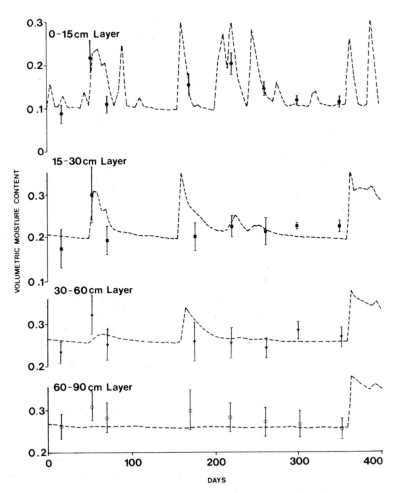

FIG. 6. Simulated volumetric soil water content of four horizons of a grey clay soil at Meandarra over a 400-day period using
the water balance model of Ross [29]. Points are field measurements
with standard deviations.

2. Simulation Using Long-Term Historical Data

One of the advantages of the discrete model of Fitzpatrick et
al. is that simulation of past soil moisture changes can be made
using historical rainfall records. Estimated days of plant growth

at four stations in central Australia using the discrete water
balance model are shown in Fig. 7. Data of this type are especially

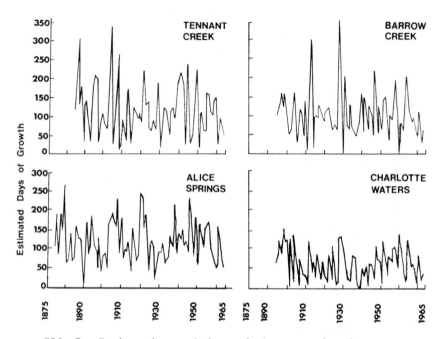

FIG. 7. Estimated annual days of plant growth using a water
balance model at four stations. Reprinted from [25] by permission.

valuable under arid and semiarid conditions and where annual variability of rainfall is high. It is also possible to derive probability distributions of climatic data and to use these as inputs to
the model instead of actual historical data.

3. Characterization of Environments

Operation of water balance models over long periods results in
the accumulation of data on estimated soil moisture levels. These
data can be used to estimate probabilities of active plant growth.
Soil moisture levels at three stations—Roseworthy (lat., 34° 33' S;
long., 138° 43' E), Taroom (lat., 25° 38' S; long., 149° 48' E), and
Townsville (lat., 19° 16' S; long., 146° 48' E) were calculated
over a period of years using a water balance model similar to that

of Fitzpatrick et al. under assumed conditions of perennial vegetation and with similar soil parameters (maximum of store A, 6.0 cm
in; store B, 2.0 cm; and apportionment of rainfall to each store in
the ratio of 75:25).

From these data, probabilities of active plant growth at 5-day
periods throughout the year could be calculated (Fig. 8). The

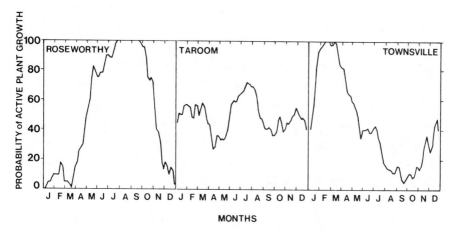

FIG. 8. Probabilities of active plant growth for three stations estimated from a water balance model.

annual evaporation (ca. 1600 mm) was similar at the three sites but
the amounts of rainfall differed (420, 685, and 1160 mm, respectively). Also, the annual distribution of both rainfall and evaporation
were different. The occurrence at two of the sites (Roseworthy and
Townsville) of periods with 100% probability of active plant growth
and of other periods with very low values is characteristic of environments with well-defined annual wet and dry periods. At Taroom,
however, there was an absence of periods of very low and very high
probability of active plant growth.

The probabilities of active plant growth shown in Fig. 8 are
based on moisture only. In effect, other factors, such as species,
day length, and temperature, are important in plant growth.
Fitzpatrick and Nix [34] proposed a multifactor growth index that
included light and thermal indices in addition to moisture and used

this to characterize environments in terms of basic biologic
responses to gross seasonal trends in climatic parameters.

4. Simulation of Plant Growth

Under many conditions of plant growth in the field it is necessary to consider the effects of limiting soil water. Water balance
models are useful in several ways in such situations. Nix and
Fitzpatrick [35], for example, used a water balance model to estimate available water supply during critical ontogenetic periods
during the growth of wheat and sorghum. They found highly significant correlations with indices calculated from these data and final
grain yields. Byrne and Tognetti [36] used a water balance model to
predict annual variation in Townsville stylo (*Stylosanthes humilis*)
dry matter yields. They estimated yields for 49 seasons and plotted
a curve of yield against frequency of occurrence. Paltridge [37]
developed a model based on the concept of limiting values, including
limiting available soil water, and their effect on plant growth
rate.

A further approach is to consider the effect of moisture level
on plant growth rate as distinct from plant yield. Richards [38,
39] reviewed the literature regarding equations for expressing plant
growth changes with time. The most comprehensive single plant
growth equation is a modification (proposed by Richards) of the von
Bertanlaffy equation. The modified equation is

$$dW/dt = [kW/(m - 1)][(A/W)^{m-1} - 1] \tag{2}$$

where W is the yield, A is the maximum yield, and k and m are parameters.

This empirical equation is a generalized one of considerable
flexibility. Although it has been fitted to plant growth under
controlled environmental conditions under which climatic parameters
remain constant throughout the growth period [38, 40] it has limitations in field situations because climatic parameters, e.g., day-
length and temperature, show sinusoidal trends.

The equation can be modified to take account of such changes as
follows

$$dW/dt = [W/(1 - m)](a_0 + a_1 t + a_2 t^2 \ldots) [(A/W)^{1-m} - 1] \quad (3)$$

or

$$dW/dt = [W/(1 - m)](a_0 + a_1 \cos t + a_2 \sin t \ldots)$$
$$[(A/W)^{1-m} - 1] \quad (4)$$

where the power series as in Eq. (3) or the Fourier series in Eq. (4) can be viewed as time-varying weighting functions.

Equations (3) and (4) do not allow for the effects of factors limiting plant growth. However, if the potential growth rate dP/dt is the maximum growth rate indicated by the genotype and environmental conditions with all nutrients freely available, then with a single limiting factor actual growth rate dW/dt is less than dP/dt and the relationship may be expressed as

$$dW/dt = (dP/dt)\{1 - \exp[-cN(t)]\} \quad (5)$$

where c is a coefficient and N(t) is a limiting factor that may be constant or time varying. The full equation may be expressed as

$$dW/dt = [W/(1 - m)](a_0 + a_1 t_1 + a_2 t^2 \ldots a_n t^n) [(A/W)^{1-m}$$
$$\{1 - \exp[-cN(t)]\} \quad (6)$$

The values of N(t), for example, could be supplied as inputs from a water balance model. It is possible to use such an equation to simulate plant growth under conditions of varying soil moisture levels. In field situations with annual plants it may be necessary to decrement the value of A and, in fact, make this time varying also.

Although Eq. (6) has a large number of parameters it is unlikely that such a complex process as plant growth under field conditions can be adequately explained with a few parameters. Also, one of the recent developments in statistics is the use of the computer in the fitting of complex models to date [41]. Programs are available that, given adequate data, can fit numbers of parameters to nonlinear regression models by minimization of least squares [42].

III. THE BIOGEOCHEMICAL CYCLE

In terrestrial ecosystems, soil is the main source of nutrients for organisms. Therefore, systems analysis of the biogeochemical cycle invariably involves consideration of the soil component.

A. Modeling of Elemental Cycles and Soil Ecosystems

Various studies have been made of the forms of chemical elements in soils. With the large number of elements involved, the complexity and variability of soils, and the existence of organomineral compounds, much of this information is, at best, rudimentary. In a systems analysis approach it is necessary to have, in addition to information concerning the static amounts of elements in compartments or pools, information of the dynamic system concerning rates of change between nutrient forms.

A great deal of data regarding nutrient cycling in vegetation and litter have been collected [43-47]. However, the emphasis has been mainly on static studies concerning the amounts of nutrients in pools. Less information on dynamic aspects is available.

Soil nutrient changes can be broadly summarized by the general equation

$$dN_i/dt = IN - OUT \tag{7}$$

where IN is the sum of all the inputs of the nutrient form N_i and OUT is the sum of all the outputs of the nutrient form.

Swartzman et al. [48] have suggested a general soil nutrient model as follows

$$dN_i/dt = f(M_i \ldots M_n, N_i, T, S, I_i) \tag{8}$$

where the principal system variables are N_i, i = 1 ... m is the concentration of the ith soil nutrient, and M_i is the active biomass of the ith microbial form. The driving functions are T, soil temperature, and S, soil moisture. The intermediate system variables is I_i, which is the sum of the inputs and outputs of the ith nutrient form resulting from all nonsoil processes.

Such a model could also be partitioned to be descriptive of various depths, in which case some account of transport of nutrient forms would be necessary.

When nutrient relationships are considered in such terms it is apparent that quantitative data on both nutrient forms and the turnover rates of nutrients are necessary if a systems analysis approach is to be meaningful. Most of this data is not available in the literature. It will have to be obtained by imaginative experimentation.

Therefore, although there is considerable potential in the application of systems analysis to nutrient cycling, little has been accomplished as yet. Most progress has been made with the movement of strontium and cesium. Interest in these elements has been associated with the radioactive contamination of soils, plants, and animals as a result of atmospheric fallout.

B. Specific Models

1. Nitrogen

Because of its important role in agricultural production a system analysis of soil-plant nitrogen interactions is desirable but is beset by difficulties. Part of the difficulties are related to the large number of forms in which nitrogen occurs and the complexity of their interactions. Hauck [49] has described recent work on some of the many forms of soil nitrogen and their interaction. Allison [50] has previously pointed out the difficulties of attempting to draw up even a crude balance sheet of soil nitrogen inputs and outputs. Dahlman and Sollins [51] used as many as 18 compartments or pools in their soil nitrogen model. A second aspect of the difficulties of a systems analysis of nitrogen in soil ecosystems is the lack of knowledge about the mechanisms and rates of change between the various forms. A third aspect is the lack of information on the amount and chemical composition of organic nitrogen, which is an important component of the nitrogen cycle. An approach to one aspect of the nitrogen cycle, namely, a mathematical

model for consecutive ammonification and nitrification reactions in
soil, has been outlined in Chapter 1, Volume 2 of *Soil Biochemistry*.

Two models will be discussed in some detail. These are the
nitrogen-water balance model of Frere et al. [52] and the grass-
legume pasture model of Ross et al. [53]. These models indicate
both the potential usefulness of a systems analysis approach to the
nitrogen cycle and the difficulty of simplifying such a complex
system.

a. <u>Nitrogen-Water Model</u>. The main processes and flows in this
model, which was programmed in FORTRAN, are shown in Fig. 9. The

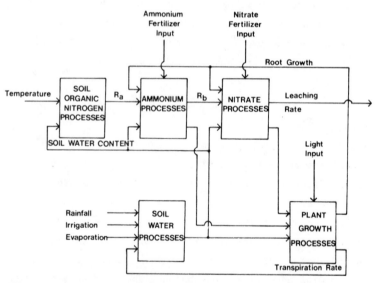

FIG. 9. Information flow diagram of water balance-nitrogen
model of Frere et al. [52]. R_a and R_b are the rates of production
of ammonium and nitrate, respectively.

specific numerical values used to test the model were for sugarbeet
grown on a Portneuf silt loam in Idaho. Nitrogen was considered to
exist in three forms in the soil: as organic nitrogen, as ammonium,
and as nitrate. The nitrate and ammonium pools were divided into
two parts: the current root zone and the remaining potential root
zone. The plant uptake rate of nitrate and ammonium in the current

root zone was calculated as being proportional to the growth rate and related to the concentration of nitrate and ammonium in the root zone.

The rate constants for ammonification and nitrification were modified for variation in soil moisture content and soil temperature. The normal curve (Fig. 10) was used as the response function,

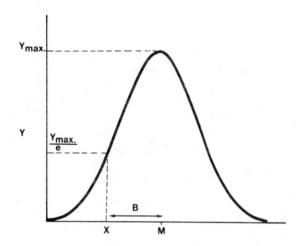

FIG. 10. Use of the normal curve as a response surface for modifying rate constants of ammonification, nitrification, and growth. (From Frere et al. [52].)

i.e.,

$$Y = Y_{max} \exp [-(M - X)^2/B^2] \tag{9}$$

For example, with temperature the optimum rate was taken at 30°C with 16°C for the B value. Ammonification or nitrification rates could then be calculated for each temperature by reference to the response function.

Some of the subsystems of this model have been validated. Others are still being subjected to sensitivity analysis. Nevertheless, some of the approaches used are of interest and suggest areas where further research is necessary on the interrelationship of components.

    b. Grass-Legume Pasture Model. Ross et al. used a systems

analysis approach to examine the effects of nitrogen and light on
the vegetative growth of grass-legume pasture. The information
flow diagram is shown in Fig. 11. The model was programmed in
FORTRAN. Carbohydrate is produced by grasses and legumes and the

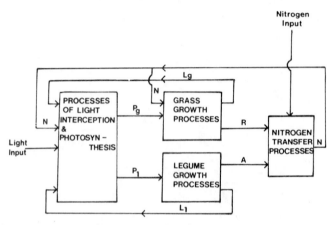

FIG. 11. Information flow diagram of grass-legume model of
Ross et al. [53]. P is the net photosynthetic rate; L is the leaf
area index (subscripts 1 and g refer to legume and grass, respectively); N is the level of available nitrogen; R is the removal rate
of available nitrogen by grass; and A is the addition rate of available nitrogen by legume to the soil.

growth processes are influenced by the level of available nitrogen.
The level of available nitrogen is, in turn, influenced by the
mineralization of soil nitrogen, by addition of nitrogen fertilizers, by transference from legume nitrogen fixed in symbiosis, and by
removal in the grass.

The model provided a satisfactory approximation of the real
system. One aspect of the model that remained unchecked was the
transfer of nitrogen through soil from the legume to the grass. The
mechanism of the transfer is unknown but it was interpreted as a
two-compartment cascade, the first compartment as nitrogen in undecomposed legume material and the second as nitrogen in material
undergoing decomposition.

The effect of nitrogen level on the behavior of the model was
examined by varying initial nitrogen levels from zero to infinity.

Daily maximum light intensity was kept constant and the run was terminated after 150 days. The effect of different levels of nitrogen on the leaf area index of grass and legume are shown in Fig. 12. Two main effects were evident; that of nitrogen on the grass component alone and that of the grass on the legume because of shading.

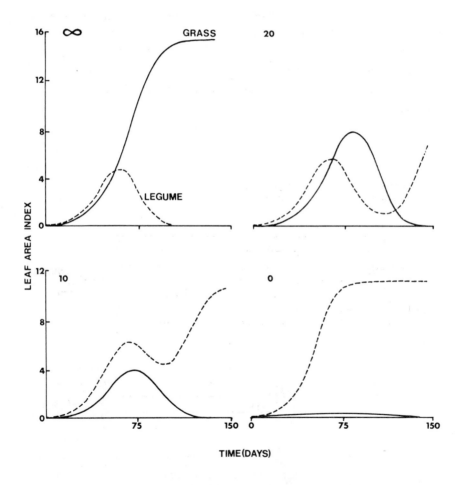

FIG. 12. Simulated effect of four different initial available nitrogen inputs (0, 10, 20, ∞) g nitrogen/m^2 on grass and legume growth (expressed as leaf area index). (From Ross et al. [53].)

The authors believe that the model can be used as a subsystem in other models provided it is restricted to the vegetative phase of growth under conditions of reasonably constant water and temperature and when water and other nutrients are nonlimiting.

2. Sulfur Model

May and Till [54-59] developed a sulfur model based on the cycling of sulfur in soil-plant-animal ecosystems. The model was programmed for an analog computer and the compartments are shown in Fig. 13. Initially, it was assumed that the transport between compartments followed first-order kinetics and no allowance was made for environmental influences. In later investigations time-varying transport rates were introduced and a better fit of the model was obtained.

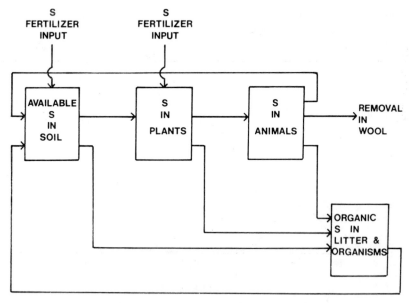

FIG. 13.   Compartment diagram of sulfur model of May and Till [54].   Boxes are quantities of sulfur; lines are rates of flow.

3. Strontium Model

Frissel et al. [60] simulated the migration of ^{90}Sr fallout on pastures and its movement through the soil using CSMP. The flow diagram of ^{90}Sr transport in the soil, which varied both in depth and time, is shown in Fig. 14 and the parameters and variables are shown in Table 2. Six rate processes were included; movement caused by mass flow, by diffusion, by biologic transport, by root adsorption, by radioactive decay, and by fixation. Two mixing processes

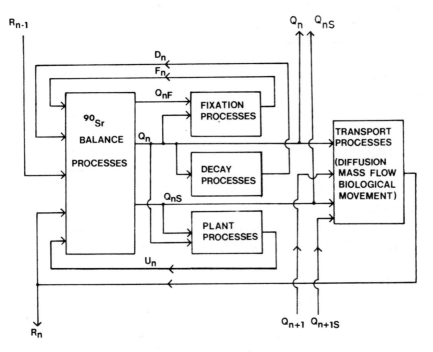

FIG. 14. Information flow diagram of the soil layer n of the ^{90}Sr model of Frissel et al. [60]. R_{n1} and R_n are the rates of movement of ^{90}Sr into and out of the layer. Q_n, Q_{nS}, and Q_{nF} are the quantities of free ^{90}Sr, soluble ^{90}Sr, and fixed ^{90}Sr, respectively, in the layer N. D_n, F_n, and U_n are the decay, fixation, and uptake rates, respectively.

TABLE 2

Parameters and variables used in the stimulation of ^{90}Sr transport
in soils. (From Frissel et al. [60].)

Parameter/variable	Dimension	Varies with
Moisture content	-	depth; time*
Moisture flux	l/t	depth; time*
Exchangeable calcium	mequiv./cm^3	depth
Ca concentration in solution	mequiv./cm^3	depth; time*
Exchange constant between Ca and Sr	-	
Ca uptake by the crop	mequiv./cm^3	time*
Uptake efficiency	-	depth
Physiological discrimination factor	-	
Diffusion constant in water	l^2/t	
Tortuosity	-	moisture content
Dispersion	l	moisture content
Initial distribution of free ^{90}Sr	mCi/km^2	depth
^{90}Sr drain rate	mCi/lm^2/t	time+
^{90}Sr drain rate reduction factor	-	time*
Initial distribution of fixed ^{90}Sr	mCi/km^2	depth
Fraction of total ^{90}Sr fixed at equilibrium	-	
Fixation rate factor	t^{-1}	
Release rate factor	t^{-1}	
Biological diffusion coefficient	l^2/t	depth
Biological mixing percentage	-	
Ploughing time	t	

* Seasonal values used repeatedly
+ Quarterly

were also included; biologic mixing and ploughing. The biologic
mixing represented the redistribution caused by worms and dead
roots. It was assumed that each year a certain fraction from each
layer was redistributed throughout the 20 layers. Ploughing was
simulated as a yearly total mixing of layers within the ploughing
depth of 20 cm.

The transport of ^{90}Sr fallout in soil has been measured since
1958 at 15 locations in the European Economic Community. Most of
the measuring sites were pastures undergoing normal grazing and soil
samples were taken yearly at 5-cm intervals down to 20 cm. Comparison of the values generated by the model and actual data at one
location (Hooglanderveen) are shown in Fig. 15. Agreement was
satisfactory. Extrapolation of the model to the near future suggests a decrease in ^{90}Sr concentrations in the surface layers of
the soil but a gradual increase in the lower layers.

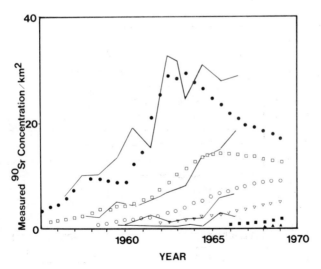

FIG. 15. Measured (—) and calculated (symbols) of ^{90}Sr
concentrations in the top layers (0-30 cm in six 5-cm layers) of a
podzol at Hooglanderveen as a function of time. Reprinted from [60]
by permission.

4. Cesium Model

Patten and Witkamp [61] used a systems analysis approach to [134]Cs kinetics. Laboratory experiments were conducted to determine patterns and rates of [134]Cs exchange in microecosystems comprised of different combinations of radioactive leaf litter, soil, microflora, millipedes, and aqueous leachate. Rate constants were determined by fitting models to data with an analog computer.

The block diagram of the system is shown in Fig. 16. The systems equations were as follows

$$dX_j/dt = \Sigma\lambda_{ij}X_i - \Sigma\lambda_{ji}X_j - \lambda_p X_j \tag{10}$$

where

X_j = radionuclide activity-density of compartment j

λ_{ij} = rate constant for the transfer of radioactive isotope from compartment i to compartment j

λ_p = rate constant for radioactive decay

t = real time (days)

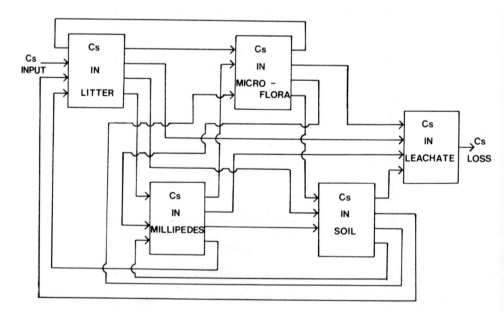

FIG. 16.   Compartment diagram of the cesium model of Patten and Witkamp [61].   Boxes are quantities of cesium; lines are rates of flow.

The rate constants for the five-compartment systems are shown in Table 3. Soil is a cesium sink at the low concentrations used in

TABLE 3

Rate Constants (per Day) of Transfer of Radioactive Cesium

| | From | | | |
To	Litter	Soil	Microflora	Millipedes
Litter	—	0	0.500	0.250
Soil	0.033	—	0	0.160
Microflora	0.120	0	—	0.250
Millipedes	0.045	0	0.250	—
Leachate	0.0009	0	0	0
Σ	0.199	0	0.750	0.660

this experiment because none of the compartments studied was capable of acquiring cesium from the soil. Later studies [62] with ^{137}Cs used combinations of compartments to separate out effects.

C. Aspects of Systems Analysis of the Biogeochemical Cycle

The main limitation in the systems analysis of the biogeochemical cycle is the lack of knowledge about the components of the system, their interactions with each other, and their rates of change. It is significant that most progress has been made in situations where isotopes provide a means of monitoring movements through the system and that the most complete model has been developed for strontium.

One requirement for more detailed systems analysis of soil ecosystems is the development of new techniques to obtain information on turnover rates. The experimental use of small watersheds in the assessment of whole ecosystems is one approach that is proving valuable [63-65]. Johnson et al. [66] studied the losses of dis- solved Ca, Mg, K, and Na from six small watersheds in the Hubbard Brook area in New Hampshire during 1963-1967. They used the formula of Barth [67] to calculate rate of chemical weathering, W, as

$$W = D_i/([c_i]-[s_i]) \qquad (11)$$

where $[c_i]$ and $[s_i]$ are the concentration of the element i in the original rock and the weathered product, respectively, and D_i is the amount of i removed in solution per unit time. Calculated weather- ing rates for the study area based on the elemental concentration in the original bedrock till surface are shown in Table 4. These data

TABLE 4

Calculated Chemical Weathering Rates in the Hubbard Brook Study[a]

Element	Element removed, D (kg/ha/yr)	Concentration of elements in:		Weathering rate of rock (kg/ha/yr)		
		Original rock, c	Weathered product, s	4-Year mean	Maximum year	Minimum year
Ca	8.0	0.014	0.004	800	1000	350
Mg	1.8	0.016	0.010	180	270	70
K	0.1	0.029	0.024	20	220	0
Na	4.6	0.011	0.001	770	1000	400

[a]Reprinted from [66] by permission.

suggest that the steady-state chemical weathering rate of the bed- rock till in this area is about 800 kg/ha/yr (for both Ca and Na). Under podzol weathering conditions a major part of the K and Mg released by the primary minerals is apparently retained by the clay and is not removed from the system.

The extension of the experimental small watershed approach to specific ecosystems should provide data on which realistic estimates of rates of change of components in the biogeochemical cycle can be made. Development of soil monitoring equipment is also necessary for more precise measurement of nutrient cycling. Techniques are

available in the case of water and salinity and specific ion electrodes appear promising for individual elements.

The development of models for elements being added in large
amounts to agricultural systems is desirable. The difficulties associated with a nitrogen model have been emphasized. Not only are
high rates of nitrogen being added as fertilizers and from automo-
bile exhausts [68] but also very high rates of nitrogen as excreta
are being added in some areas by the bioindustries [69]. The
ultimate effects of these changes in nitrogen balance cannot be
foreseen. Clearly, much more basic work on soil nitrogen transformations is necessary, particularly to supply data for models and
to indicate the major transformations and their quantitative relationships.

There has been some interest in phosphorus changes in lakes
and streams [70] as a consequence of eutrophication [71]. Less attention has been given to a systems analysis approach to soil phosphorus. Large amounts of phosphorus are added as fertilizers and
also considerable amounts are removed in agricultural products.
Many time-varying changes in soil phosphorus have been noted. The
buildup of residual phosphorus levels [72-75], the gradual change
of added forms of phosphorus [76], and the effects of soil properties on rate of plant uptake [77] all suggest that a systems analysis approach may be useful.

A general model for transformations in the soil of cationic
nutrients may be possible. Large amounts of potassium are being
added to agricultural systems and such a model would be valuable.
Radford [4] has presented an outline of a plant-soil model for potassium.

IV. ENERGY FLOW

Unlike the water and biogeochemical cycles, which are true
cycles, the energy cascade represents a continuous input of solar

energy and its flow and gradual dissipation throughout the ecosystem. The soil plays a less direct role than in the water and biogeochemical cycles, yet aspects of energy flow are of extreme importance to the general stability of soil ecosystems. The steadystate level of soil organic matter in any ecosystem depends on an equilibrium between production and decomposition processes. Aspects of the efficiency of energy conversion are particularly important in organic matter balance.

A.   Modeling of Energy Flow and Soil Ecosystems

Various approaches have been made to modeling higher plant growth and physiological processes. Models of the effect of light on plant growth and photosynthesis and respiration have been developed [78-85]. Such processes as transpiration [86], stomatal closure [87, 88], and root growth have also been modeled [89]. Soil interactions are absent in many of these models. Even in models where some soil interaction occurs, such as in the ELCROSS model of Brouwer and De Wit [89], in which root temperature is an input that effects the relative growth rate of roots, soil effects are only a minor part of the model.

Soil organic matter represents a major compartment in the lower stages of the energy flow cascade (Fig. 3) and transformations of soil organic matter are of intrinsic interest in agricultural ecosystems. Trends in soil organic matter levels provide guidelines to the stability of the ecosystem and the permanence of the management practices being employed. If either the rate of plant production is decreased or the rate of soil organic matter decomposition is increased, soil organic matter levels will be decreased and a new and lower equilibrium level is likely, which may influence such other soil properties as structure.

Various mathematical models have been applied to organic matter change in soil. Much of the earlier work on soil organic matter trends was a simple systems analysis approach. However, the models were generally a single equation only. Soil organic matter content has usually been expressed in terms of organic nitrogen or organic carbon.

Jenny [90] proposed that soil organic matter change with time could be expressed as a first-order reaction

$$dN/dt = - K_1 N + K_2 \tag{12}$$

where

N = soil organic matter expressed as nitrogen or carbon

K_1 = rate of decomposition

K_2 = rate of addition

Various modifications of these equations have been made [91-94]. One change to further generalize the equation is to consider the rate coefficients as seasonal and expressed as Fourier coefficients [95], i.e.

$$dN/dt = - (\tfrac{1}{2}\alpha_0 + \alpha_1 \cos t + \beta_1 \sin t \ldots)N + (\tfrac{1}{2}\gamma_0 + \gamma_1 \cos t + \phi_1 \sin T \ldots) \tag{13}$$

Attempts to fit models of soil organic matter to real data have been made in a wide range of agricultural systems [96-99].

The systems analysis approach allows more sophisticated models to be developed. In particular, various effects, such as improvement of plant genotypes and fertilizer usage on the increased efficiency of energy conversion, can be considered.

B. Specific Model: Carbon-14

Neel and Olson [100] developed a model of the movement of organic carbon and [^{14}C] in a terrestrial ecosystem. The model was programmed for an analog computer and the block diagram is shown in Fig. 17.

The four main compartments in the model were the organic carbon amounts in vegetation, roots, organic litter, and soil. These compartments were numbered 1, 2, 4, and 6 respectively.

The main system equations were

$$dV/dt = P_1 - k_1 V \tag{14}$$

$$dR/dt = \phi_{12} k_1 V - k_2 R \tag{15}$$

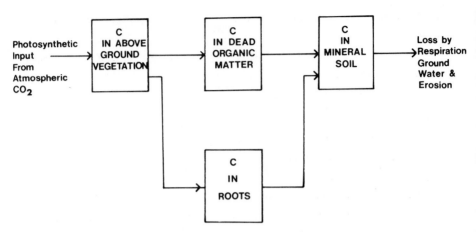

FIG. 17. Compartment diagram of the ^{14}C model of Neel and Olson [100].

$$dD/dt = \phi_{14}k_1V - k_4D \tag{16}$$

$$dS/dt = \phi_{46}k_4D - k_6S + k_2R \tag{17}$$

Where V is above-ground vegetation, P is the inflow of carbon by photosynthesis, R is roots, D is dead organic matter, and S is mineral soil. k_1, k_2, k_4, and k_6 represent decay parameters of compartments 1, 2, 4, and 6, respectively. ϕ_{12}, ϕ_{14}, and ϕ_{46} represent partial transfer coefficients. Behavior of the model was examined under four conditions of varying energy flow: (a) constant, (b) oscillating, (c) decreasing sinusoidal, and (d) increasing sinusoidal.

The above model was used to simulate the distribution of ^{14}C in the ecosystem. Broecker and Olson [101] estimated that nuclear weapons testing has increased the overall natural equilibrium of ^{14}C in the earth's atmosphere. They projected future concentrations in the atmosphere, assuming no further tests. A pulse potential simulating the estimated changes in atmospheric $^{14}CO_2$ over a 50-year period was fed into the model. Vegetation showed a prompt increase followed by a fairly prompt decrease to normal levels. The soil compartment showed the greatest delay and the greatest attenuation in percentage increase of ^{14}C. However, the slow decay parameter for this compartment also resulted in a larger increase in total ^{14}C than for the litter compartment. Moreover, the soil compartment still retained its excess ^{14}C long after ^{14}C in other compartments had come and gone.

C. Aspects of Energy Flow Models

At the present time, systems analysis is being increasingly applied to plant mechanisms affecting energy flow. In addition to the limited number of papers cited in this chapter, other aspects are being examined [102]. Less research is being applied to soil organic matter changes and to the interaction of plant yields and soil organic matter levels, although some studies have been made [103]. Much experimental data on yield trends and soil changes in agricultural systems are still being accumulated. Experiments vary from classical studies, such as those that have been in progress for 125 years at Rothamsted [104], to the measurement of changes in new environments brought about forest removal and cultivation in the semiarid subtropics [105].

There have been significant changes in the efficiency of energy conversion in many agricultural systems. In the United States, average corn yields after remaining constant at 1500 kg/ha for nearly a century have increased to 5000 kg/ha in recent years [109]. Auer [106, 107] analyzed the sources of yield improvement over the period 1939-1960 and found that 69% could be attributed to two pro- cesses--genetic improvement and fertilizer application--and that the effects of these were of the same magnitude.

Therefore, any model of plant yield trends and soil organic matter change quantitative expression of genetic change is neces- sary. A general equation of genetic change may be considered as

$$dY/dt = z(t)(P - Y) \qquad (18)$$

where Y_t is the potential yield of the genotype at time t, P is the ultimate potential yield, and $z(t)$ is a time-varying coefficient. In specific situations in which the same genotype and technology are used for long periods, z(t) may be zero. Under the conditions of genetic change that have occurred in recent years a more realistic approach may be to express changes in terms of the three-parameter logistic or autocatalytic functions, i.e.

$$Y_t = P/(1 + be^{-kt})$$ (19)

where P is the ultimate potential yield and b and k are parameters.

Systems analysis also allows the feedback effects of yield on soil organic matter to be simulated by equations of the form

$$dN/dt = -K_1N + K_2 + K_3Y(t)$$ (20)

where Y(t) represents dry matter yield in the ecosystem.

The relatively slow change of soil organic matter with time presents experimental difficulties. Half-lives of soil organic accumulation and decay are of the order of 10-10^2 years and field experimental approaches are, at best, slow and time consuming. Systems analysis may be of value in making use of available information, in understanding the system, in predicting future trends, and in suggesting further soil and plant parameters that should be measured.

V. CONCLUSIONS

In this chapter some published examples of the systems analysis approach to soil ecosystems have been given. The models discussed have been of diverse but usually simple systems. The fragmentary nature of the models is undoubtedly a reflection of the difficulties of analyzing soil ecosystems and the state of the art of model building generally. Model building of biologic systems is in its very early stages and evolutionary changes are rapid creating difficulties in publication and communication, both for the analyst and for the scientific journals generally. Some of the slow development of the systems analysis approach in soil ecosystems as compared with other fields are a result of the complexity and variability of soils and of past studies of soil scientists concentrating on the static rather than the dynamic aspects of soil entities.

In one sense, as De Wit [108] has pointed out, an analyst of biologic systems is confronted with a complex and intricate system of many levels, including molecules, cell structures, cells, tissues, organs, individuals, populations, and communities. At best, only small sections of this system are known. Success in systems analysis is most likely if quantitative and dynamic analyses are restricted to the simplest ecosystems.

Experimentally, the systems analysis approach is likely to have an effect on the type of research investigations undertaken and the data collected. There is likely to be more investigation into mechanisms and processes generally, more emphasis on rates rather than yields, and more logical thought applied to the interaction of systems variables.

It can already be predicted that multistage models, that is, models connecting several different levels of knowledge, will be beyond the capacity of present day computers. However, current limitations of the application of systems analysis to soil ecosystems do not lie with computer technology. Rather, the limitations are in the knowledge of the mechanisms of soil ecosystems and the lack of quantitative information of soil and plant parameters. Much can be accomplished with the hardware currently available. Also, evolutionary change in digital computers in relation to size, speed, and cost per calculation is still occurring.

There are likely to be three main benefits from the application of systems analysis to soil ecosystems. The first benefit arises from the discipline involved in thinking about the whole system and the interactions of its entities and in attempting to simplify its complexities. The second benefit comes from the exercise of constructing the model. It is true that a validated model tells us little that was not previously known because the behavior of the model was defined during construction. However, the necessity for expressing qualitative information in quantitative terms, the exposure of areas of inadequate information, the need for a conceptual framework, and the possible need to suggest mechanisms not thought of previously are all by-products of the systems analysis approach.

The third benefit arises from the operation of the validated model. This often increases understanding of the ecosystem and allows manipulation of variables in ways that may not be possible experimentally. Operation of the model may also allow extrapolation into areas not studied and prediction of future behavior. At the very least the approach allows hypothesis generation, which is at the core of the scientific method.

REFERENCES

1. D. T. Ross, *Proc. 22nd Natl. Conf. Assoc. Comput. Mach.*, p. 367, Thompson Book Co., Washington, 1967.
2. M. B. Dale, *Ecology*, *51*, 2 (1970).
3. L. C. Chapas, in *The Use of Models in Agricultural and Biological Research* (J. G. W. Jones, Ed.), p. 72, Grassland Res. Inst., Hurley, 1969.
4. P. J. Radford, in *The Use of Models in Agricultural and Biological Research* (J. G. W. Jones, Ed.), p. 87, Grassland Res. Inst., Hurley, 1969.
5. R. D. Brennan and M. Y. Silberberg, *IBM Syst. J.*, *6*, 242 (1967).
6. R. D. Brennan, C. T. De Wit, W. A. Williams, and E. V. Quattrin, *Oecologia*, 4, 113 (1970).
7. G. M. Van Dyne, *Colo. State Univ. Range Sci. Dept. Sci. Ser. No. 3*, 1969.
8. P. J. Ross, CSIRO, Div. Soils Tech. Paper No. 20 (1973).
9. J. R. Emshoff and R. L. Sisson, *Design and Use of Computer Simulation Models*, Macmillan, New York, 1970.
10. A. B. Clymer, *Proc. Conf. Appl. Contin. Syst. Simul. Lang.*, p. 1, San Francisco, 1969.
11. J. H. Milsum, *Biological Control Systems Analysis*, McGraw-Hill, New York, 1966.
12. P. R. Benyon, *Simulation*, *11*, 219 (1968).
13. A. G. Tansley, *Ecology*, *16*, 284 (1935).
14. E. G. Knox, *Soil Sci. Soc. Amer. Proc.*, *30*, 79 (1965).
15. J. S. Russell and A. W. Moore, *Trans. 9th Int. Cong. Soil Sci.*, *4*, 205 (1968).
16. A. W. Moore, J. S. Russell, and W. T. Ward, *J. Soil Sci.*, *23*, 193 (1972).
17. C. T. De Wit and H. Van Keulen, *Simulation of Transport Processes in Soils*, Div. Theor. Prod. Ecol., Wageningen, 1970.
18. P. J. Wierenga and C. T. De Wit, *Soil Sci. Soc. Amer. Proc.*, *34*, 845 (1970).
19. E. Bresler, *Soil Sci.*, *104*, 227 (1967).

20. W. Baier, *Inst. J. Biometerol.*, *9*, 5 (1965).
21. M. M. Hufschmidt and M. B. Fiering, *Simulation Techniques for the Design of Water Resource Systems*, MacMillan, New York, 1966.
22. W. C. Boughton, *Univ. New South Wales Water Resources Lab. Report*, *78*, Sydney, 1965.
23. W. C. Boughton, *J. Hydrol. (New Zealand)*, *7*, 75 (1968).
24. D. D. Huff, D. G. Watts, O. L. Loucks, and M. Teraguchi, *IBP Deciduous Forest Biome Memo Report*, *No. 70-4* (1970).
25. E. A. Fitzpatrick, R. O. Slatyer, and A. I. Krishnan, *Agri. Meterol.*, *4*, 389 (1967).
26. W. Baier and G. W. Robertson, *Agri. Meteorol.*, *5*, 17 (1968).
27. W. Baier, *Agri. Meteorol.*, *6*, 151 (1969).
28. J. McAlpine, *Proc. 11th Int. Grassland Cong.*, p. 484, 1970.
29. P. J. Ross, *CSIRO, Div. Soils Techn. Mem. No.*, 22/71 (1971).
30. S. T. Willatt, *Agrl. Meteorol.*, *8*, 341 (1971).
31. E. A. Fitzpatrick, *J. Appl. Meteorol.*, *2*, 780 (1963).
32. R. Winkworth, *Agri. Meteorol.*, *7*, 387 (1970).
33. R. L. McCown, *Austral. J. Exp. Agri. Anim. Husb.*, *11*, 343 (1971).
34. E. A. Fitzpatrick and H. A. Nix, in *Australian Grasslands* (R. M. Moore, Ed.), A.N.U., Canberra, 1970.
35. H. A. Nix and E. A. Fitzpatrick, *Agri. Meteorol.*, *6*, 321 (1969).
36. G. F. Byrne and K. Tognetti, *Agri. Meteorol.*, *6*, 151 (1969).
37. G. W. Paltridge, *Agri. Meteorol.*, *7*, 93 (1970).
38. F. J. Richards, *J. Exp. Bot.*, *29*, 290 (1959).
39. F. J. Richards, in *Plant Physiology* (F. C. Steward, Ed.), Vol. 5A, Chap. 1, Academic, New York, 1969.
40. D. J. C. Friend, V. A. Helson, and J. E. Fisher, *Can. J. Bot.*, *40*, 939 (1962).
41. K. E. F. Watt, *Ecology and Resource Management: A Quantitative Approach*, p. 242, McGraw-Hill, New York, 1968.
42. P. J. Ross, *CSIRO, Div. Soils Techn. Paper No. 6*, 1971.
43. J. D. Ovington, *Advan. Ecol. Res.*, *1*, 103 (1962).
44. L. E. Rodin and N. I. Bazilevich, *Production and Mineral Cycling in Terrestrial Vegetation* (Transl. G. E. Fogg), Oliver and Boyd, London, 1967.
45. P. Duvigneaud and S. Denaeuer De-Smet, *Ecol. Stud. Anal. Syn.*, *1*, 200 (1970).
46. A. W. Moore, J. S. Russell, and J. E. Coaldrake, *Austral. J. Bot.*, *15*, 11 (1967).
47. J. S. Russell, A. W. Moore, and J. E. Coaldrake, *Austral. J. Bot.*, *15*, 481 (1967).
48. G. Swartzman et al., *IBP Grassland Biome. Tech. Report No. 32*, Fort Collins, Colorado, 1970.
49. R. D. Hauck, *Trans. 9th Int. Congr. Soil Sci.*, *2*, 475 (1968).
50. F. E. Allison, *Advan. Agron.*, *7*, 213 (1955).
51. R. C. Dahlman and P. Sollins, *Agron. Abstr.*, Amer. Soc. Agron. Meeting, Tucson, 1970.

52. M. H. Frere, M. E. Jensen, and J. N. Carter, *Proc. 1970 Summ. Comput. Simul. Conf.*, Denver, p. 746.
53. P. J. Ross, E. F. Henzell, and D. R. Ross, *J. Appl. Ecol.*, *9*, 535 (1972).
54. P. F. May, A. R. Till, and A. M. Downes, *Austral. J. Agri. Res.*, *19*, 531 (1968).
55. A. R. Till and P. F. May, *Austral. J. Agri. Res.*, *21*, 253 (1970).
56. A. R. Till and P. F. May, *Austral. J. Agri. Res.*, *21*, 455 (1970).
57. A. R. Till and P. F. May, *Austral. J. Agri. Res.*, *22*, 391 (1971).
58. P. F. May and A. R. Till, *Austral. Plant Nutr. Conf.*, Section 1(d), p. 49, Mt. Gambier, S. Australia, 1970.
59. A. R. Till, P. F. May, and I. W. McDonald, *Symposium on Sulphur in Nutrition*, Chap. 12, Avi, Wesport, Conn., 1970.
60. M. J. Frissel, P. Poelstra, and P. Reiniger in *Soc. Chem. Ind. London Monograph*, *37*, 135 (1970).
61. B. C. Patten and M. Witkamp, *Ecology*, *48*, 813 (1967).
62. M. Witkamp and M. L. Frank, in *Symp. Radioecol.*, Proc. 2nd Natl. Symp., p. 635, Ann Arbor, 1967.
63. F. H. Borman and G. E. Likens, *Science*, *155*, 424 (1967).
64. N. M. Johnson, G. E. Likens, F. H. Borman, D. W. Fisher, and R. S. Pierce, *Water Resources Res.*, *5*, 1353 (1969).
65. N. M. Johnson, *Ecology*, *52*, 529 (1971).
66. N. M. Johnson, G. E. Likens, F. H. Borman, and R. S. Pierce, *Geochim. Cosmochim. Acta*, *32*, 531 (1968).
67. T. F. W. Barth, *Geochim. Cosmochim. Acta*, *14*, 204 (1961).
68. B. Commoner in *Global Effects of Environmental Pollution* (S. F. Singer, Ed.), p. 70, D. Reidel, Dordrecht-Holland, 1970.
69. D. R. Keeney and W. R. Gardner, in *Global Effects of Environmental Pollution* (S. F. Singer, Ed.), p. 96, D. Reidel, Dordrecht-Holland, 1970.
70. A. D. Hasler, in *Global Effects of Environmental Pollution* (S. F. Singer, Ed.), p. 110, D. Reidel, Dordrecht-Holland, 1970.
71. A. M. Beeton, *Limnol. Oceanog.*, *10*, 240 (1965).
72. K. Woodrooffe and C. H. Williams, *Austral. J. Agri. Res.*, *4*, 127 (1953).
73. J. S. Russell, *Austral. J. Agri. Res.*, *11*, 927 (1960).
74. C. S. Piper and M. P. C. De Vries, *Austral. J. Agri. Res.*, *15*, (1964).
75. E. A. Jackson, *CSIRO, Bull.*, No. 284 (1966).
76. S. Larsen, D. Gunary, and C. D. Sutton, *J. Soil Sci.*, *16*, 141 (1965).
77. N. J. Barrow, *Soil Sci.*, *104*, 99 (1967).
78. W. G. Duncan, R. S. Loomis, W. A. Williams, and R. Hanan, *Hilgardia*, *38*, 187 (1967).
79. W. G. Duncan, W. A. Williams, and R. S. Loomis, *Crop Sci.*, *7*, 37 (1967).

80. W. A. Williams and R. S. Loomis, *Winter Meeting Amer. Soc. Agri. Eng.*, Paper 69-913, Chicago, 1969.

81. W. G. Duncan, *Proc. Conf. Appl. Contin. Syst. Simul. Lang.*, p. 137, San Francisco, 1969.

82. R. S. Loomis, G. W. Fick, and W. A. Williams, *Proc. Conf. Appl. Contin. Syst. Simul. Lang.*, p. 137, San Francisco, 1969.

83. C. T. De Wit and R. Brouwer, *Z. Angewan. Bot.*, *42*, 1 (1968).

84. D. W. Stewart and E. R. Lemon, *Microclimate Investigations*, Interim Report, 69-3, USDA and Cornell University, Res. and Dev. Tech. Report, ECOM 2-68 1-6 (1969).

85. C. T. De Wit, R. Brouwer, and F. W. T. Penning de Vries, *Proc. IBP/PP Tech. Meeting*, p. 47, Trebon, 1969.

86. I. Evangelisti and R. Van der Weert, *A Simulation Model for Transpiration of Crops*, Report No. 2, Dept. Theor. Prod. Ecol., Wageningen, 1971.

87. K. B. Woo, L. N. Stone, and L. Boersma, *Water Resources Res.*, *2*, 71 (1966).

88. K. B. Woo, L. N. Stone, and L. Boersma, *Water Resources Res.*, *2*, 85 (1966).

89. R. Brouwer and C. T. De Wit, in *Root Growth* (W. J. Whittington, ed.), p. 224, Butterworths, London, 1969.

90. H. Jenny, *Factors of Soil Formation*, McGraw-Hill, New York, 1941.

91. J. E. Dawson, in *Agricultural Chemistry* (D. E. H. Frear, Ed.), Vol. 1, Van Nostrand, New York, 1950.

92. C. M. Woodruff, *Soil Sci. Soc. Amer. Proc.*, *14*, 208 (1949).

93. S. Henin, G. Monnier, and L. Turc, *Compt. Rend. Acad. Sci. Paris*, *248*, 138 (1959).

94. W. V. Bartholomew and D. Kirkman, *Trans. 7th Int. Cong. Soil Sci.*, *2*, 471 (1960).

95. J. S. Russell, *Nature (London)*, *204*, 161 (1964).

96. P. H. Nye and D. J. Greenland, *Commonwealth Agri. Bureau Tech. Comm.*, *No. 51* (1960).

97. H. Laudelot, *Dynamics of Tropical Soils in Relation to Their Fallowing Techniques*, FAO, Rome, 1960.

98. J. S. Russell, *Austral. J. Agri. Res.*, *11*, 902 (1960).

99. D. J. Greenland, *Soils Fert.*, *34*, 237 (1971).

100. R. B. Neel and J. S. Olson, *Use of Analog Computers for Simulating the Movement of Isotopes in Ecological Systems*, Oak Ridge National Laboratory, ORNL-3172, 1962.

101. M. W. Broecker and E. A. Olson, *Science*, *130*, 712 (1960).

102. Anon., Prediction and Measurement of Photosynthetic Activity, *Proc. IBP/PP Tech. Meeting*, Trebon, Centre for Agricultural Publishing and Documentation, Wageningen, 1969.

103. R. C. Dahlman and C. L. Kucera, in *Symp. Radioecol.*, Proc. 2nd Natl. Symp., p. 652, Ann Arbor, 1967.

104. Anon., *Details of the Classical and Long Term Experiments up to 1967*, Roth. Experimental Station, Harpenden, 1970.

105. Annual Report, 1968-69, p. 33, CSIRO, Div. Trop. Past., St. Lucia, Queensland, 1969.

106. L. Auer, Ph.D. Thesis, Iowa State University, Ames, 1963.
107. L. Auer, *Econ. Council Canada Staff Study, No. 24*, 1969.
108. C. T. De Wit, *Proc. IBP/PP Tech. Meeting*, p. 17, Trebon,
     Centre for Agricultural Publishing and Documentation
     Wageningen, 1969.
109. J. S. Russell, *J. Austral. Inst. Agri. Sci. 39*, 156 (1973).

CHAPTER 3

MICROBIAL METABOLISM OF FLOODED SOILS

Tomio Yoshida

The International Rice Research Institute
Los Baños, Laguna, Philippines

I. INTRODUCTION

Soil is a complex ecosystem containing a wide variety of microorganisms. These organisms bring about a large number of biochemical changes. In flooded soils the changes are so varied and numerous that it is difficult to bring the processes together into

a single unified system. The main biochemical processes in flooded
soils, however, can be regarded as a series of successive oxidation-reduction reactions mediated by different types of bacteria.
Flooding alters the character of the microbial flora in soils.
Generally, aerobes are replaced by facultative anaerobes, which in
turn are superseded by anaerobes. The successive microbial changes
are accompanied by a stepwise biochemical and chemical reduction
of the soil, a lowering of oxidation-reduction potential (Eh), and
a change in pH. These changes may mask soil differences to some
extent and also influence the growth of plants in flooded soils.
Because rice is the most important crop grown on flooded soils,
this review is largely confined to flooded rice soils.

   This chapter deals with the following aspects of the microbial
metabolism of flooded soils: (1) development of the milieu of
flooded soils; (2) aerobic respiration; (3) anaerobic respiration;
(4) fermentation; (5) degradation of insecticides; and (6) the
rhizosphere of rice.

II. DEVELOPMENT OF THE MILIEU OF FLOODED RICE SOILS

   The operations involved in the wet cultivation of rice pro-
foundly influence microbiological activity in the soil. The usual
operations include (1) submergence of the soil, with or without
puddling, for the duration of the crop, with or without soil drying
in midseason; (2) draining and drying the soils before harvest; and
(3) reflooding for the next crop a few weeks to several months
after harvest. Superimposed on these are inherent properties of
the soil or environment that affect the microbiological regime in
rice soils.

   Several studies of microflora in paddy soils in Japan [1-3],
India [4, 5], and Egypt [6] indicate that bacteria are the predominant microorganisms in flooded rice soils. Aerobic bacteria reach
their peak numbers a few days after a soil is flooded. The peak
numbers of facultatively anaerobic bacteria and anaerobic bacteria

follow [7].

Takai et al. [8, 9] have divided the reducing process of paddy soil into two main stages: one before and the other after the reduction of iron is completed. In the first stage they observed evolution of CO_2, absorption of O_2, and the attainment of the peak bacterial numbers. At the same time nitrate disappeared, manganic and ferric ions were reduced and soil Eh dropped rapidly. Ammonia and CO_2 were liberated as reduction progressed. The second step is anaerobic. Organic acids, CH_4, H_2, and H_2S were produced and populations of anaerobic bacteria increased. Within one or few days after the soil had flooded, O_2 disappeared; during the next 4-9 days, CO_2 was evolved and accumulated. After CO_2 accumulation reached a maximum it began to decrease and CH_4 evolved concurrently with H_2. The amount of organic acids reached a maximum within 2-9 days of incubation and then markedly decreased. Takai and Kamura [10, 11] developed a scheme of the successive progress of microbial metabolism in the flooded soil (Table 1).

TABLE 1

Successive Microbial Reduction Processes in Flooded Soils [10]

Transformation of elements	Initial soil Eh	Biochemical pattern
Disappearance of NO_3^-	+ 0.6 $\sim$ + 0.5	Aerobic respiration
Disappearance of NO_3	+ 0.6 $\sim$ + 0.5	Anaerobic respiration
Formation of Mn^{2+}	+ 0.6 $\sim$ + 0.4	Anaerobic respiration
Formation of Fe^{2+}	+ 0.6 $\sim$ + 0.3	Anaerobic respiration
Formation of S^{2-}	0 $\sim$ - 0.19	Anaerobic respiration
Formation of H_2	- 0.15 $\sim$ - 0.22	Fermentation
Formation of CH_4	- 0.15 $\sim$ - 0.19	Fermentation

In most biological systems, organisms oxidize a compound by dehydrogenation, which is usually a stepwise removal of two hydrogen atoms and two electrons from the molecule. The biological oxidation-reduction reaction involved in transporting hydrogen through the electron-transporting system provides the ultimate source of energy. In aerobic respiration, molecular oxygen is used as a terminal acceptor of hydrogen atoms and electrons. In the absence of molecular oxygen, other oxidized soil components can function as electron acceptors. Microorganisms transfer the produced energy during the oxidation-reduction reaction mostly into adenosine triphosphate (ATP) molecules and use the free energy stored in ATP for their biosynthesis. The efficient use of free energy stored differs depending on the species of bacteria and the growth conditions.

Takai [12] calculated the free energy change at pH 7 (Delta F') for each bacterial function shown in Table 1 on the basis of the available data on chemical changes in the flooded soils. In this scheme, bacteria in flooded soils produce more energy through aerobic respiration. The free energy change was -56.7 kcal for aerobic respiration, -53.6 kcal for denitrification by *Pseudomonas denitrificans*, -9.6 kcal for sulfate reduction by *Desulfovibrio desulfuricans*, -9.2 kcal for the methan fermentation by *Methanobacterium omelianskii* and -6.7 kcal by *Methanosarcina barkeri*. The biochemical functions in flooded soil proceed from higher-energy-yielding reactions to lower-energy-yielding reactions. The molar growth yields (Y_{ATP}) in the step that uses the stored energy for synthesis of bacterial cells, do not differ much among aerobic, facultatively anaerobic, and anaerobic heterotrophs; the average value of cells was 10.5 g/mole of ATP generated [13].

After flooding, when reduction processes in a flooded soil reach a steady state, the soil becomes a biochemical environment quite different from upland soil. The metabolic activity of the soil microorganisms weakens, and the oxygen supply from the atmosphere and oxygen released by algae and aquatic weeds, through photosynthesis, exceeds the amount of oxygen consumed by the soil microorganisms in the surface soil. As a result, an oxidizing horizon, called the "oxidized layer," develops in the upper part of the flooded soil [14, 15]. The surface of the soil where the molecular oxygen is available remains oxidized and supports aerobic metabolism

by bacteria. Nitrification can take place in the oxidized soil
[14-17].

Beneath the oxidized soil surface layer, the soil is anaerobic
and reduced. This is the root zone of rice. Anaerobic respiration
and fermentation occur mostly in the reduced soil layer. In contrast to the yellowish brown oxidized layer, the reduced layer is
bluish gray probably because of the presence of certain ferrous
compounds [14, 15].

III. AEROBIC RESPIRATION IN FLOODED SOIL

Like the energy needs of other living organisms, those of the
soil microbes are met by respiration, with oxidation of organic compounds and liberation of free energy. In aerobic biological oxidation, molecular oxygen is the ultimate electron acceptor. In the
presence of molecular oxygen, organic substrates are oxidized mainly
to CO_2, whereas molecular oxygen is reduced to water.

The aerobic heterotrophs actively metabolize the organic matter
in soil shortly after the flooding. The amount of oxygen required
by microorganisms in soil at that time may exceed the amount supplied by diffusion through the overlying water. Molecular oxygen
in the soil is consumed within a few days after flooding [7, 18, 19].
When most of the easily decomposable organic matter is consumed, the
supply of oxygen in the surface soil exceeds the demand for oxygen
by soil bacteria. The flooded soil then develops an oxidized layer
at the surface. The concentration of dissolved oxygen in flood water during the day varies from 2 to 18 ppm, depending on the time
it is measured, and it sometimes reaches air saturation or higher
[19]. However, Patrick and Sturgis [20] found no oxygen 1 cm below
the soil-water interface of flooded soils. Carbohydrates for the
heterotrophic bacteria in the oxidized layer of flooded soil can be
supplied from the anaerobic zone in the reduced layer by diffusion.

A theoretical model [21] suggested that consumption of molecular oxygen in the oxidized layer can be increased by the diffusion of the products of anaerobic respiration in the reduced layer to the aerobic zone. The oxygen concentration that limits the aerobic respiration of bacteria is 1 % of the air-saturated level under standard temperature and pressure [22]. The molecular oxygen available in flood water, therefore, does not seem to limit the activity of aerobic bacteria in the surface soil where molecular oxygen is available. The biological activity of aerobic bacteria can also occur in the surface of soil, but the aerobic zone may be limited to a fairly thin surface soil layer and may require a considerable flux of oxygen to maintain the oxidized state [21].

The root-soil boundary of flooded soils planted to rice may also be a possible site for aerobic bacteria. Rice plants and other aquatic plants diffuse O_2 from roots to the surrounding media [23, 24]. Because the total surface area of plant roots is much larger than the soil surface occupied by the plants [25], the root-soil boundary in flooded soil may be an important area for aerobic metabolism if the plant provides the rhizosphere bacteria with enough molecular oxygen to permit aerobic respiration.

Another group of bacteria, i.e., autotrophic bacteria, are mostly aerobic and use molecular oxygen as their terminal electron acceptor. They derive their energy by oxidizing inorganic compounds and depend mostly on CO_2 fixation as their carbon source. These are the nitrifying bacteria, ferrous iron-oxidizing bacteria, sulfur-oxidizing bacteria, methane-oxidizing bacteria, or hydrogen-oxidizing bacteria. The biochemistry of autotrophic metabolism by the bacteria has been reviewed elsewhere [26].

The occurrence of nitrification in flooded soil has been suggested by Shioiri [14] and others [15, 16]. If ammonium fertilizer is applied to the surface layer of flooded soil that is in an oxidized state, the ammonium is oxidized to nitrate by nitrifying bacteria. If the nitrate is subsequently translocated to the reduced layers, it becomes susceptible to denitrification. Recovery of nitrogen from ammonium fertilizers by rice plants is better when nitrogen is applied to the reduced layer than when it is applied to

the surface soil in flooded fields [15, 27-34]. These studies suggest that nitrification takes place in the oxidized layer and that
subsequent denitrification takes place in the reduced layer of
flooded soils. However, few biochemical studies have been made on
nitrification in flooded soils. Manzano [35] and Tusneem [36] found
that considerable nitrogenwas lost in open systems in which gas
diffused freely to the atmosphere, but that losses were negligible
when soils were incubated in a closed system or in anaerobic conditions. It was recently demonstrated by using isotopic nitrogen
that N_2 that evolved from soil is derived from ammonium applied to
flooded soil and that no nitrogen was lost when oxygen was excluded
[37]. The results show that nitrification in the oxidized zone was
very active and that denitrification occurred simultaneously in the
soil system.

Motomura [38,39] studied the mechanism involved in the formation
of manganese sediments, often found in Japanese rice soils and
reported that bacteria play an important role in manganese oxidation.
The bacterial oxidation of ferrous irons can take place in the oxidized layer or in the rhizosphere where the aerobic iron bacteria
can meet ferrous iron as well as free oxygen. The iron bacteria,
which depend for their energy source on the oxidation of ferrous
iron to ferric iron at pH values near 7, must compete with chemical
oxidation since the oxidation of ferrous iron by oxygen at this pH
is very rapid [40].

Methane-oxidizing bacteria, possibly present in the oxidized
layer of flooded soils, can oxidize methane to carbon dioxide [41]
or it may be used anaerobically by photosynthetic bacteria [42].
Molecular hydrogen is aerobically oxidized to water by H_2-oxidizing
bacteria.

Autotrophic bacteria of the genus *Thiobacillus* can oxidize
sulfur compounds, mainly sulfides in flooded soils, to sulfate in
aerobic zones. The oxidized rhizosphere of the rice plant may make
sulfur available to the flooded rice plant. In many rice-growing
countries, rice is grown on acid sulfate soils. These soils occur
in marine flood plains and are derived from parent materials bearing

FeS_2 (pyrite) [43]. When a flooded acid sulfate soil is drained, the sulfides are oxidized. The sulfur-oxidizing bacteria play an important role in forming sulfate and in lowering the pH of acid sulfate soils [44, 45]. Sulfides can be also oxidized chemically in the aerobic zones of flooded soil [46-48].

IV. ANAEROBIC RESPIRATION IN FLOODED SOILS

Anaerobic respiration is a biological, energy-yielding, oxidation-reduction reaction in which an inorganic compound other than oxygen is used as the external electron acceptor. When the aerobic microorganisms have used up the oxygen in soil after flooding, anaerobic respiration by facultatively anaerobic bacteria becomes dominant. The facultatively anaerobic bacteria use nitrate, manganic oxides, ferric oxides, sulfate, carbonate, and perhaps other moieties as the electron acceptors and reduce them to molecular nitrogen, manganous and ferrous compounds, sulfide, methane, and other reduction products. Mineral reduction by anaerobic respiration is coupled with energy-yielding oxidation of organic compounds and, occasionally, or inorganic compounds by the bacteria.

A. Nitrate Reduction

The nitrate ion produced in the aerobic zones or derived from applied nitrate fertilizer (not commonly used in rice culture) is readily susceptible to reduction by various nitrate-reducing bacteria in flooded soils when the supply of molecular oxygen becomes limited. The nitrogen-reducing systems of microorganisms are classified into three categories: assimilation, respiration or dissimilation, and N_2 fixation. Nitrogen respiration is further divided into three kinds of biochemical functions: that for accumulating nitrite, that for accumulating ammonia, and that for denitrifying to nitrogen and nitrous oxide gases. Nitrate reduction by denitrifying bacteria can take place in the presence of molecular oxygen if the nitrate concentration is low, both in media [49, 50] and in soils [51]. In the absence of oxygen the facultative anaerobes actively use nitrate as an electron acceptor during oxidation of carbohydrate. In glucose or acetic acid oxidation by denitrifying bacteria, nitrate is converted to nitrogen gas:

$$C_6H_{12}O_6 + 4\ NO_3^- \longrightarrow 6\ CO_2 + 6\ H_2O + 2\ N_2 \tag{1}$$

$$5\ CH_3COOH + 8\ NO_3^- \longrightarrow 10\ CO_2 + 6\ H_2O + 8\ OH^- + 4\ N_2 \tag{2}$$

MacRae et al. [52] were able to recover some of the applied ^{15}N from the organic fractions of soils 6 weeks after submergence and only a small fraction of the applied ^{15}N was recovered as available ammonia [52]. Because they observed a considerable loss of ^{15}N-labeled nitrate that was added to flooded soils the major nitrate reduction process was probably denitrification. The denitrifying activity of soil is not necessarily related to the organic matter content of flooded soil [52, 53]. Autotrophic denitrification by *Thiobacillus denitrificans* and *Micrococcus denitrificans*, which can oxidize sulfide or hydrogen, respectively, anaerobically in the presence of nitrate, may have a role in denitrification in flooded soils. The sulfur-oxidizing bacteria denitrify nitrate as follows:

$$5\ S + 6\ NO_3^- + 2\ H_2O \longrightarrow 5\ SO_4^{2-} + 3\ N_2 + 4\ H^+ \tag{3}$$

A common characteristic of denitrifiers is that they produce nitrous oxide N_2O in addition to N_2 [54]. Nitrous oxide is present in minute amounts in the atmosphere. Arnold [55] reported that the soils appeared to be a possible source of the gas to the atmosphere. Under certain soil conditions, the gas can be a major initial product in denitrification [56]. A denitrifying bacterium produces abundant nitrous oxide when there is abundant nitrate [57]. Nitrifying bacteria also produce the gas during oxidation of ammonia to nitrite under certain conditions [58, 59]. The formation of nitrous oxide in flooded soil during nitrate reduction or during nitrification by soil bacteria has not been reported.

Some *Bacillus sp.* isolated from flooded soils accumulate nitrite in pure culture with nitrate [60]. Nitrite does not usually accumulate in flooded soils, however. No report has shown that an excess of nitrite or ammonia accumulates in flooded neutral soils during nitrogen respiration of nitrate.

B. Manganese Reduction

At pH 7.0 manganese dioxide can function as an electron acceptor in anaerobic respiration at an oxidation level comparable to that of nitrate. Microbial reduction of manganese dioxide can take place in the presence of hydrogen donors and manganese dioxide can act as a terminal hydrogen acceptor under anaerobic conditions [61, 62]. Apparently soil bacteria are involved in the reduction since glucose stimulates the reduction of manganese and the effect of the added glucose is retarded by adding azide in the soil [61]. The bacteria that reduce manganese dioxide have been isolated, but few biochemical studies of manganese reduction have been made [63, 64]. The addition of organic matter to flooded soils increase the manganous ion in soil [65, 66]. Manganese reduction is illustrated by

$$CH_3COOH + MnO_2 \longrightarrow 2 CO_2 + Mn^{2+} + 4 H^+ \qquad (4)$$

The presence of nitrate in flooded soils retards manganese reduction [11], suggesting that the facultatively anaerobic bacteria uses nitrate preferentially as an electron acceptor.

C. Iron Reduction

About the time of maximum accumulation of reduced manganese compounds in flooded soil, the amount of ferrous iron starts to increase [11]. Bromfield [67] isolated bacteria from soil that actively reduce ferric iron to ferrous iron under anaerobic conditions in the presence of oxidizable substances. The formation of ferrous iron in flooded soil is largely associated with microbiological

functions [68-70]. As with manganese reduction, the addition of
organic matter to the soil enhances the reduction of iron [65, 66,
70], but the presence of nitrate or of MnO_2 in the soils retards
it [11].

In experiments by Asami and Takai [71], the amount of iron
reduced was highly correlated with the amount of carbon dioxide
produced for about 2 weeks after flooding. They suggested that carbon dioxide evolves as a result of organic acid degradation by bacteria when ferric iron is used as an electron acceptor:

$$CH_3COOH + 8\ Fe^{3+} + 2\ H_2O \longrightarrow 2\ CO_2 + 8\ Fe^{2+} + 8\ H^+ \tag{5}$$

Because the amount of iron in soils is usually much higher
than the nitrate or manganese the transformation of iron greatly
influences the pH and the Eh of the soil [12, 70, 72-74]. The gley-
formation in flooded soil is partly caused by iron reduction by
bacteria [67, 70, 75-77]. Ferric oxides may act as hydrogen acceptors for clostridia in flooded soils [77].

Although microbes probably play a role in iron reduction in
flooded soil, few studies indicate that minerals are actually involved directly as electron acceptors for microbial respiration.
Some aerobic microorganisms that possess nitrogen reductase reduce
ferric iron to ferrous iron [78, 79]. The addition of nitrate and
chlorate but not of sulfate has been found to decrease the iron-
reducing activity of the bacteria [80]. By using a cell-free extract of a marine *Bacillus* sp., DeCastro and Ehrlich [81] demonstrated the iron-reducing activity of the bacterium. However, the
bacterium does not possess a nitrate reductase system.

D. Sulfate Reduction

The amount of sulfide accumulated in flooded soils increases
remarkably at a soil Eh of about -200 mV [11, 66, 82]. The anaerobic bacterium, *Desulfovibrio desulfuricans*, is probably a major
group in the respiratory sulfate reduction in flooded soils [46,
83]. This bacterium possesses cytochrome c_3 which has low redox

potential (E_o = -204 mV) and requires a low $\hat{E}$h for its growth [84].
The sulfate-reducing bacteria can grow both autotrophically and
heterotrophically:

$$4 \ H_2 + SO_4^{2-} \longrightarrow S^{2-} + 4 \ H_2O \tag{6}$$

$$2 \ CH_3CHOHCOOH + SO_4^{2-} \longrightarrow 2 \ CH_3COOH + 2 \ H_2O + 2 \ CO_2 + S^{2-} \tag{7}$$

The biochemistry of the bacterial metabolism has been reviewed
elsewhere [26, 85, 86]. The presence of nitrate inhibits sulfate
reduction, and sulfide forms only after all the nitrate has disappeared from the flooded soil [47].

Manganese dioxide apparently cannot compete with sulfate as a
hydrogen acceptor but can oxidize hydrogen sulfide once it forms
[83]. In old, degraded paddy soils, called "Akiochi" soils in Japan,
H_2S that accumulates in the soil is believed to be toxic to the
rice plant [15]. The amount of ferrous iron in Akiochi soils is
about one-fifth that in normal paddy soils [87]. The amount of
active iron in soils is correlated with the production of free H_2S,
probably because the active iron forms insoluble iron sulfide in
soil [87, 88].

E. Carbon Dioxide Reduction

Carbon dioxide is susceptible to anaerobic CO_2 respiration by
methane bacteria. Carbon dioxide produced by the oxidation of organic matter or present in flooded soils is reduced to methane by
anaerobic respiration, although the biochemical reduction mechanism
has not been established. Several bacteria are capable of using
molecular H_2 in reducing CO_2 to CH_4:

$$CO_2 + 4 \ H_2 \longrightarrow CH_4 + 2 \ H_2O \tag{8}$$

Although H_2 is a common end product of anaerobic metabolism, loss of
H_2 to the atmosphere from flooded soils is small because CH_4-producing
bacteria use molecular H_2 as a source of energy for growth [41].
Takai [89] found that the proportion of CH_4 derived from CO_2 added

to flooded soils was 10-15 % in one soil and 25 % in a soil that contained more organic matter.

F. Reduction of Other Elements

Tsubota [90] detected phosphine gas, an end product of phosphate reduction by dephosphorification, in flooded soils to which organic matters and ammonium phosphate had been added. He also detected the metabolic intermediates, phosphite and hypophosphite, in the phosphate reduction in a medium inoculated with *Clostridium butyricum*, which he isolated from the soil.

The oxidation-reduction reaction between iodide and molecular iodine can probably take place in flooded soils [91, 92]. In flooded soil more iodine is available to the rice plant than in upland soil. Addition of glucose to the flooded soil greatly increases the amount of available iodine, suggesting that a biochemical reduction system for the element was present in flooded soils.

A *Pseudomonas* isolated from soil fixes large amounts of mercury or mercuric chloride or phenylmercuric acetate on the cell surface [93]. The bacterium is able to reduce the mercury ion anaerobically in the compounds to the metallic form [94].

$$Hg^{2+} + 2 e^- \longrightarrow Hg \tag{9}$$
$$CH_3Hg^+ + H^+ + e^- \longrightarrow CH_4 + Hg \tag{10}$$

Various inorganic compounds are biochemically reduced with molecular hydrogen by *Micrococcus lactilyticus* [95].

V. FERMENTATION IN FLOODED SOIL

Anaerobic bacteria in flooded soils degrade carbohydrates by a process called "fermentation." "Fermentation" is defined as a biological, energy-yielding, oxidation-reduction reaction in which organic compounds serve as the final electron acceptors. The biochemical processes of fermentation by bacteria have been summarized by others [26, 96, 97]. Pyruvate, a terminal metabolite of glycolysis in anaerobic processes, is a key compound in the biochemical pathway of anaerobic carbohydrate breakdown. The pattern of fermentation differs depending on the species of bacteria. The main end products of fermentation by bacteria are ethanol, formate, acetate, lactate, propionate, butyrate, molecular hydrogen, and carbon dioxide [26, 97].

Fermentation of nitrogenous compounds, or putrefaction, produces ammonium, amines, indole, skatole, mercaptans, and hydrogen sulfide as well as the end products carbon dioxide and hydrogen, in addition to organic acids and alcohols [98]. The organic acids and alcohols eventually are converted to methane and carbon dioxide in flooded soils.

A. Organic Matter Fermentation

Since 1905, many workers have reported that organic acids are produced in flooded rice soils [99]. Organic acids in rice paddy soils have been widely studied because they inhibit the growth of the rice plant [100]. There are many reports on the production of aliphatic organic acids in flooded rice soils [8, 101-111]. The studies of organic acids in flooded soils have been reviewed by Stevenson [112]. Among the aliphatic acids, lower volatile fatty acids -- formic, acetic, propionic, and butyric acids -- and lactic acid are often reported as fermentation products in flooded soils. Pyruvic acid, which is produced mainly as a terminal compound in the anaerobic degradation of glucose by the Embden-Meyerhof pathway, is a key intermediate in the production of lower fatty acids.

The reaction products of pyruvate depend on the bacteria acting on it. The biochemical pathways of each fermentation have been established [26, 96]. In flooded soil some of the fermentation patterns probably take place simultaneously during anaerobic degradation of organic matter. Anaerobic bacteria, clostridia, are distributed abundantly in flooded soils. Their major fermentation products are acetic, butyric, and lactic acids in addition to carbon dioxide and hydrogen [26, 96, 113]. A *clostridium* ferments lactic acid to propionic acid with acrylate as an intermediate [96].

According to Takai [12], in high-yielding paddy soils, acetic acid starts to accumulate within 10-14 days after flooding; in normal rice soils, within 6-8 days; in degraded Akiochi soils within one day. The amounts of acetic acid at the highest peaks were 0.2-0.7 m.e. per 100 g in high-yielding paddy soils, 0.4-0.6 m.e. per 100 g in normal rice soils, and 0.9-1.4 m.e. per 100 g in degraded Akiochi soil. The degraded Akiochi soil accumulates butyric acid rapidly in the early period after flooding. A small amount of butyric acid was found in the high-yielding paddy soils, however. Sato and Yamane [107] found that additions of glucose, starch, cellulose, and gelatin to a flooded soil increased the formation of acetic and butyric acids. The velocity of acid production decreased in this order: gluclose > starch > gelatin > cellulose.

Data on organic acid accumulation, measured by the continuous flow technique, showed that glucose anaerobically fermented to acetic, butyric, formic, and lactic acids [114]. The level of organic acids decreased remarkably 10 days after the soil was incubated. The addition of nitrate to anaerobic soils reduces the amount of volatile lower fatty acids that accumulate [115]. Probably, in the anaerobic respiration of nitrate, denitrifying bacteria convert the acids to carbon dioxide. The nitrate ion may also inhibit the methane-producing bacteria, which use the acids as substrate [116]. Wang and Chuang [117] reported that more lower aliphatic alcohols, particularly n-butanol, were produced in flooded soils to which green manure was added than in aerobic soil.

Anaerobic decomposition of amino acids, purines, pyrimidines, and other nitrogenous compounds by bacteria has been reviewed by Barker [118]. About 20 species of bacteria are known to ferment single amino acids and eight species of bacteria are known to ferment heterocyclic nitrogen compound. These species belong mostly to the genera *Clostridium* and *Micrococcus*. The fermentation of nitrogenous organic compounds in flooded soils has not been studied, but it is not difficult to visualize that some of the fermentation products in flooded soil reported in the literature are derived from various nitrogenous organic compounds.

Addition of fresh leaves of *Crotalaria juncea* to flooded soils produces the aromatic acids p-hydroxybenzoic, vanillic, and p-coumaric [110]. The phenolic acids were probably released from the aromatic constituents of green leaves during decomposition. The continual incorporation of such organic materials as weeds, rice straw, and root residues in flooded soil which generally is oxygen poor should favor accumulation of organic materials that contain lignin, tannin, and other aromatic compounds [119].

Microbial degradation of aromatic compounds, including hydroxylation and subsequent ring fission, requires oxygen [120]. Although anaerobic degradation of aromatic compounds in flooded soil has not been studied, some studies on the anaerobic degradation of aromatic compounds by microorganisms have been made [121-124]. A pseudomonad metabolizes benzoate and p-hydroxybenzoate anaerobically in the presence of nitrate and takes a biodegradation pathway distinct from known aerobic metabolism [124]. Benzoic acid is probably a key intermediate in the anaerobic metabolism of aromatic compounds whereas in aerobic metabolism protocatechuate is a key intermediate [124]. The anaerobic metabolic pathway in the degradation of p-hydroxybenzoic acid by the pseudomonad through benzoic acid has not been elucidated.

Anaerobic degradations of aromatic compounds by photosynthetic bacteria under light have been reported [125-127]. Anaerobic metabolism of benzoic acid by the photosynthetic bacterium, *Rhodopseudomonas palustris*, is a novel reductive pathway. Dutton and Evans [125] proposed that benzoic acid changes anaerobically to pimelic acid through intermediates such as cyclohex-1-ene-1-carboxylate, 2-hydroxycyclohexane carboxylate, and 2-oxycyclohexane carboxylate. The anaerobic fermentation of aromatic compounds has also been suggested [128]. It involves the dehydrogenation of the aromatic ring and formation of alicyclic intermediate compounds followed by a cleavage of the ring structure to form methane. A study on slow and incomplete anaerobic degradation of lignin in corn stalk has been reported [129]. This has not been confirmed by other workers [130].

B. Gas Metabolism

Carbon dioxide and N_2, which evolve rapidly soon after flooding, are largely the result of aerobic or anaerobic respiration of carbohydrates and of nitrate respiration if nitrate is present in the soil. They are not end products of fermentation. Carbon dioxide and H_2 are the typical gas compounds in carbohydrate fermentation. Harrison and Aiyer [131], however, found large amounts of CH_4 and smaller amounts of H_2 in flooded soil. Yamane and Sato [132] could not find a significant amount of H_2 in flooded soils except in soils to which glucose had been applied. The H_2 evolved and disappeared rapidly in the glucose-added soil. Probably H_2 was used by methane bacteria as an energy source to produce CH_4 [133]. Hydrogen evolves as formic acids decrease, suggesting that H_2 is derived from the formic acid. However, the addition of formic acid to flooded soil did not result in significant evolution of H_2 [133]. An experiment with $^{14}CO_2$ showed that CO_2 reduction may be caused not only by CH_4 formation but also by reduction of CO_2 to acetic acids [89]. Methane gas starts to evolve soon after the lower fatty acids and H_2 have started to decrease in the flooded soils [107, 133] or when CO_2 decreases in flooded soils [8, 11]. The biochemistry of methane formation by bacteria has been reviewed elsewhere [26, 41, 96, 134-136].

The various mechanisms of methane formation depend on the kind of bacteria and substrates. The methane bacteria are generally restricted to the use of relatively simple organic and inorganic compounds [134]. In flooded soil two major biochemical pathways produce CH_4: One is the reduction of CO_2 with H_2 or organic molecule as a hydrogen donor, and the other is the simple decarboxylation of acetic acid. With ^{14}C tracer techniques, Takai [89] recently showed that the methane formed in flooded soils is derived largely by transmethylation of acetic acid (decarboxylation) and to some extent by reduction of CO_2. The biochemical mechanism of methane formation from acetic acid has been studied in detail [137-140].

Although methanol and ethanol change to methane directly or indirectly after oxidation to organic acids [135], the alcohols have not often been reported as intermediates in fermentation in flooded soil. The methane fermentation from propionic acid can be significant in flooded soils if the concentration is high, although the biochemical mechanism of methane formation from propionic acid has not been well established [135]. However, Yamane and Sato [141] found that the addition of propionic acid to a flooded soil did not cause much methane formation compared with the additions of formic, acetic, butyric, or lactic acid. Most CH_4 in flooded soil is probably produced by acetic acid fermentation [89, 142].

Molecular nitrogen is the major gas component in soil immediately after flooding, but the addition of organic material decreases the ratio of N_2 gas in the gas phase as CO_2 and CH_4 evolve [143]. According to Bell [143], however, in unamended soil 92% of the N_2 gas was present in the flooded soil 98 days after flooding and 49% was present 126 days after flooding. Most of the gas phase was occupied by CH_4 and CO_2 in 16 days and 48 days of flooding when peptone and cellulose were added, respectively. The major gas components in a flooded rice field during a rice growing season are CH_4 and N_2, according to Harrison and Aiyer [131]. Interestingly at the later growth stages of rice plant more than 70% of the gas phase in flooded soils planted to rice is nitrogen, whereas about 35% of the gas phase in unplanted paddy soils was nitrogen [131].

Addition of nitrate decreases the accumulation of CH_4 in anaerobically incubated soils [144]. Methane formation in flooded soils was also retarded by adding ferric iron [71]. These may be explained by the competition of CH_4-producing bacteria with the denitrifying bacteria or the iron-reducing bacteria for volatile fatty acids as substrate. Figure 1 shows the major organic acid and methane fermentation of carbohydrates in flooded soil.

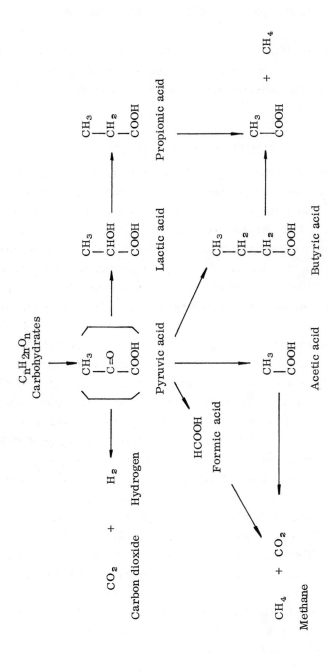

FIG. 1. A proposed diagram showing major organic acid and methane fermentation of carbohydrates in flooded soils.

VI. DEGRADATION OF INSECTICIDES IN FLOODED SOIL

Most studies of insecticide degradation in soil have been made in aerobic soil. The persistence and the biodegradation pathway of insecticides in flooded soils differ substantially from those in upland soils.

A. Organochlorine Insecticides

Many organochlorine insecticides persist for a long time in upland soils [119, 145] and consequently in upland crops [146, 147]. However, in flooded soils microorganisms rapidly degrade γ-BHC (γ-isomer of 1,2,3,4,5,6-hexachlorocyclohexane) [148, 149]. Yoshida and Castro found that very little of γ-BHC added was degraded under upland soils, whereas much of the γ-BHC was degraded in a month in flooded soils [150]. The extent of γ-BHC degradation in the soils depended on the organic matter content of soil. The higher the content of organic matter, the faster the degradation of the insecticide. The addition of nitrate of manganic dioxide, but not sulfate, retarded the rate of γ-BHC degradation [150]. A bacterium that degrades γ-BHC anaerobically has been isolated from a flooded soil [151]. The biodegradation of γ-BHC is inhibited in the presence of molecular oxygen or nitrate [152]. Although the biochemical pathway of the γ-BHC degradation is not well elucidated, the first intermediate of anaerobic degradation seems to be γ-pentachlorocyclohex-1-ane ($C_6H_7Cl_5$), not γ-pentachlorocyclohex-1-ene ($C_6H_5Cl_5$) [152]. The γ-pentachlorocyclohex-1-ene (γ-PCCHene) has been identified in metabolites of γ-BHC and is considered as the first intermediate of γ-BHC degradation in houseflies [153-155], in rats [156], or in soil [157]. Because large amounts of γ-PCCHene have not been found in γ-BHC-treated houseflies, Bridges [155] suggested that direct dechlorination, not dehydrochlorination, is the first step in the pathway of γ-BHC metabolism.

The biochemical function that converts DDT [1,1,1-trichloro-2,2-bis(p-chlorophenyl)ethane] to DDD [1,1-dichloro-2,2-bis(p-chlorophenyl) ethane] has been called reductive dechlorination; one chlorine atom in a molecule is replaced by one hydrogen atom. The reductive dechlorination of DDT to DDD has been reported by many workers [158-162]. Because the bacterium that degrades γ-BHC converts DDT to DDD anaerobically [152], the bacterium isolated from paddy soil may convert γ-BHC by reductive dechlorination to the first intermediate, γ-pentachlorocyclohexane (γ-PCCHane).

More active γ-BHC degradation occurs in anaerobic than in aerobic waste-water sludge [163]. In an simulated anaerobic impoundment, γ-BHC degraded much faster than in an aerobic impoundment [164]. The two peaks in the gas chromatograms of metabolite products produced during γ-BHC degradation, and the retention times of the compounds were identical to those of α-BHC and δ-BHC, suggesting a biological isomerization of BHC [164]. Ninety percent of added γ-BHC seemed to isomerize to α- and δ-BHC under anaerobic condition in the simulated lake impoundments [164]. No bacteria that are able to isomerize γ-BHC in nature have been isolated. All isomers α, β, γ, and δ of hexachlorocyclohexane are degraded in flooded soils with a short lag period, with the β and δ forms in the early stages of incubation [165].

In flooded soils, DDT converts rapidly to DDD [166]. The reductive dechlorination of DDT under flooded conditions was more active in soil with higher organic matter content [166]. Although the reductive dechlorination of DDT to DDD can occur in higher animals [167] and microorganisms [152, 158, 160, 161, 168-171], the dechlorination of DDT to DDD in higher animals requires the presence of molecular oxygen whereas in microorganisms the presence of molecular oxygen hinders the dechlorination [170]. Stenersen [169] found that the facultative anaerobes converted 90% of added DDT to DDD and only 5% to DDE [1,1-dichloro-2,2-bis(p-chlorophenyl)ethylene] under anaerobic conditions. In bacterial metabolism of DDT, DDE, which is the dehydrochlorination product of DDT, is not likely to be the first intermediate [161]. Wedemeyer [170] reported that reduced cytochrome oxidase is probably the cellular agent in the reductive dechlorination of DDT to DDD by *aerobacter aerogenes*.

The conversion of DDT to DDD seems most likely to occur in anaerobic environments because it takes place in anaerobic soils

[166, 172-175], in anaerobic sludge [163], in anaerobic lake impoundments [164], in rumen fluid, and in lake water [176]. In studies by
Guenzi and Beard [173] 62% of DDT added to soil was converted to DDD
under anaerobic conditions at $30^{\circ}C$. The rate of conversion of DDT
to DDD is inversely related to the O_2 concentration in soil. As with
γ-BHC, the degradation of DDT is enhanced by the application of organic matter to soil [172, 175]. Methoxychlor [1,1,1-trichloro-2,2-
bis(p-methoxyphenyl)ethane], which possesses the methoxyl group in
the benzene moiety of DDT, also degrades in flooded soil at a rate
similar to that of DDT [166]. DME [1,1-dichloro-2,2-bis(p-methoxyphenyl)ethane], as a reductive dechlorinated derivitive of methoxychlor, is produced by a culture of *A. aerogenes,* which reduces DDT
to DDD [177].

There are reports of microbial degradation of cyclodiene insecticides [158, 178-181]. Very few studies are available, however, on
the degradation of cyclodienes in anaerobic environments [163, 166].
The anaerobic degradation of the cyclodiene insecticides, heptachlor
[1,4,5,6,7,8,8-heptachloro-3a,4,7,7a-tetrahydro-4,7-methanoindene]
occurs also in flooded soils [166]. Heptachlor undergoes epoxidation
to yield heptachlor epoxide in soil [182]. Miles et al. [180] reported
that chlordane was produced by soil bacteria during the dechlorination of heptachlor. An intermediate of heptachlor in anaerobic environment may not be the result of the epoxidation function. Castro
and Yoshida [166] did not detect heptachlor epoxide in flooded soils
to which heptachlor had been applied. In anaerobic sludge, heptachlor converts quickly to an unidentified product that is more persistent than heptachlor [163]. Because the soil bacterium isolated
from the flooded soil can degrade methoxychlor and heptachlor as
well as γ-BHC under anaerobic conditions [175], the bacterial enzyme
is probably common to these insecticides in carrying out the reductive dechlorination. The first biochemical reactions proposed in
the biodegradation of organochlorine insecticides mentioned above
are summarized in Fig. 2.

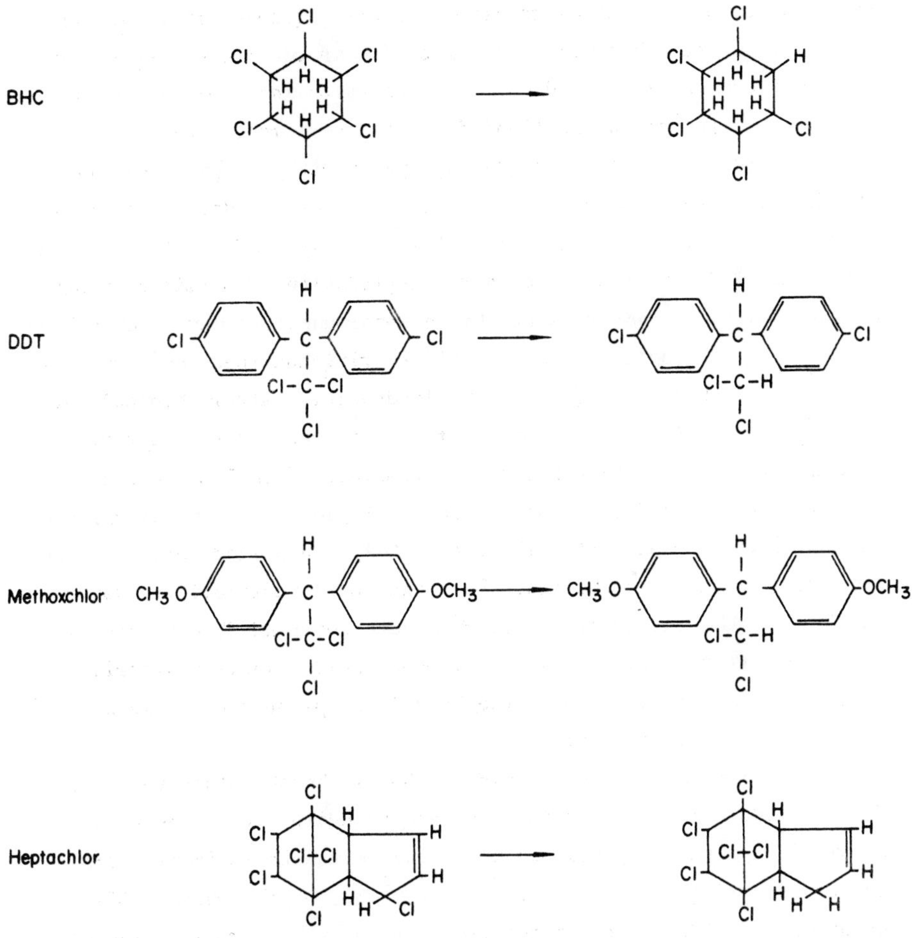

Fig. 2. The proposed first steps in the biodegradation pathway
of some organochlorine insecticides in flooded soils.

B. Organophosphorus Insecticides

Many organophosphorus insecticides are more soluble in water
and more susceptible to hydrolysis than are organochlorine insecticides. The organophosphorus insecticides are considered relatively
nonpersistent in plants and soils. Reports on the degradation of
this type of insecticde in flooded soils are scarce, however.

Diazinon [o,o-diethyl-o-(2-isopropyl-6-methyl-4-pyrimidinyl)
phosphorothioate] has been used in flooded rice fields. Early literature on diazinon residues in soil was mainly confined to nonflooded
conditions. The initial step in the degradation of diazinon in nonflooded soils is hydrolysis at the phosphorous-oxygen-pyrimidine bond,
followed by disruption of the pyrimidine ring and the subsequent release of carbon dioxide [183]. In flooded soils after chemical and
biological hydrolysis, the amount of an intermediate (2-isopropyl-
6-methyl-4-hydroxypyrimidine) that accumulated from ^{14}C-labeled
diazinon was about 43% in the natural soil and 20% in the sterilized
soil during the month after flooding [184]. In upland soils, during
the same period of incubation, 8% of the total radioactivity was recovered as the intermediate from fumigated soil and 2% was recovered
from nonfumigated soil [183]. In flooded soils, where most soil
microorganisms are anaerobic, the hydrolysis product may remain
longer than in upland soil.

Parathion (o,o-diethyl-o-p-nitrophenyl phosphorothioate) is used
for pest control in agricultural crops including rice. The major
pathway of parathion breakdown by bacteria in lake sediments [185],
in rumen fluid [186], by yeast in soil [187], by rhizobium [188] or
by *chlorella* [189] involves the reduction of the nitro-N in the molecule to produce the first intermediate, aminoparathion. Applied
parathion degraded rapidly in flooded soils but persisted in nonflooded soils [190]; Aminoparathion was detected in the flooded soils.
The anaerobic conditions in flooded soils probably enhances the reduction of the nitro-nitrogen in the molecule to amino-nitrogen.

VII. THE RHIZOSPHERE OF RICE IN FLOODED SOIL

The root zone of the plant is the most important area of the plant-soil relationship. Biochemical changes in flooded soil influence the plant through that region. Plants also influence the flooded soil through their biochemical activities in that region which is commonly called the "rhizosphere." The rhizosphere has been defined by various workers [191-193]. In this discussion, "rhizosphere" refers to the soil-plant root interface, including the surface of root tissues and the surrounding soil. Microbiological and biochemical studies of the rhizosphere have been reviewed by Rovira and Mc-Dougall [194], but few studies of the rhizosphere in flooded soil have been made.

A. Oxidative and Reductive Character of the Rhizosphere

Root tissues of aquatic plants generally possess an intercellular space under excessive moisture conditions that apparently transports air passed through stomata [195]. Rice plants grown in flooded soil have a unique biochemical characteristic. The ability of rice roots to function efficiently in anaerobic conditions is one of the most interesting and important aspects of rice culture. Some workers have shown that the rice plant supplies molecular oxygen to the roots [23, 196-199]. The gas space in the roots of rice plants constitutes 5-30% of root tissue, whereas the gas space in roots of barley is less than 1% [200]. The air-transporting system of the rice plant, which consists of the aerenchyma and the lysigenous intercellular space, develops to a greater extent under flooded condition than under upland conditions [201, 202].

Rice roots and other aquatic plants can oxidize even a portion of the rhizosphere [15, 23, 203]. The ability to oxidize the rhizosphere, according to Bouldin [203], has important implications because it alleviates the effect of toxic reduced products, it affects phosphorus nutrition, and it generates CO_2 by aerobic oxidation of organic materials in the rhizosphere.

The oxidative activity of roots is predominantly high in seedlings of lowland rice (rice varieties grown in flooded soils) in contrast to upland rice [15]. Despite the oxidative characteristic of rice roots the rice rhizosphere tends to be more reductive at the later growth stages than at the earlier growth stage of plant [15]. The nitrate-reducing activities, determined by the amount of nitrate formation in water culture of rice with nitrate, start early in the reproductive phase (panicle initiation, heading, and flowering stages) of the rice plant, reaches a maximum during the heading to the flowering stages, and decrease thereafter [204].

Nitrate reduction increased remarkably when the top of the plant was detached from the root but was not affected by the light exposure [205]. Any factor that retards the air-transporting activity of the rhizosphere may enhance the reduction state of the rice rhizosphere. The respiration rate of rice roots, determined by oxygen consumption, shows a maximum at about the heading stage of rice plant [206, 207]. It appears that the diffusion of oxygen to the rhizosphere decreases when the respiratory activity of roots become high.

The reductive condition of the rice rhizosphere, at the reproductive phase of rice plant growth, is enhanced by an increase in organic materials from root exudates or dead cells of root hairs [15]. The root exudates from a variety of plants have been listed by Rovira [208]. Little information is available on the chemical composition of the rice root exudates [209, 210]. The amino acids reported in the exudates are lysine, asparagine, aspartic acid, threonine, serine, glutamic acid, glycine, alanine, methionine, isoleucine, tyrosine, phenylalanine, cystine, and other unidentified amino acids. The carbohydrate components in the exudates of rice seedlings are glucose, fructose, arabinose, xylose, and sucrose. Because it is difficult to grow rice plants under aseptic conditions beyond the seedling stage, data have not been obtained on the chemical components or rice root exudates for the whole growth period of the plant. The amount of carbohydrates in the rice culture solution, as determined by the permanganate consumption method, indicates that

the quantities of organic matter in the rice rhizosphere are higher
in the later stages of growth than in the earlier stages [204]. However, it is not known whether this is attributable to the root exudates or to root debris. Reduced organic compounds produced under
anaerobic conditions in flooded soils also may diffuse to the rice
rhizosphere. These organic materials increase microbial activities
and more oxygen is consumed in the rhizosphere.

There are no data on the extent of the rice rhizosphere in
flooded soils. The extent of the rhizosphere in which the oxidizing
activity of roots operates can be very limited. But the number and
the amount of rice roots rapidly increase during the vegatative
stages of plant growth and reach a maximum at the heading stage [207].
According to Alberda [211] the superficial root mat, in which the
roots are thin and abundantly branched, starts to develop at the end
of the tillering stage and lasts until the ripening stage. The
oxidized area in the rhizosphere of rice may become appreciable in
the later stages of rice growth.

B. Mineral Transformations

It is difficult to differentiate between the role of root tissue
and the role of bacteria in transformations of chemicals in the
rhizosphere, because the bacteria are closely associated with the
root tissue. Although some workers have suggested that the rice
plant itself is able to fix atmospheric nitrogen [212], others dis-
agree [213]. It is difficult to determine the small amount of nitrogen that might be fixed in the plant. However, with the acetylene
reduction method, it was demonstrated that the rice rhizosphere can
fix atmospheric nitrogen [214,215]. The fixation of atmospheric
nitrogen in rice roots was actually found in bacteria that inhabit
the rhizophere rather than in the plant tissue itself [214]. Nitrogen fixation in paddy soils is much higher in fields in which rice
is growing than in fallow fields and it is higher in flooded
soils than in upland soils [175, 190]. Dommergues et al. [215]
reported that the nitrogenase activity in rice rhizosphere responds

greatly to light intensity. Döbereiner and Ruschel [216] reported a successful inoculation of an aerobic N_2-fixing bacterium, *Beijerinckia*, in the rhizosphere of rice under flooded soil conditions. These reports indicate the close relationship between N_2-fixing activity and plant growth in the rice rhizosphere.

If the rice plant is able to transport air to the rhizosphere, atmospheric nitrogen must also be transported to the roots without difficulty.

Carbohydrates are available as the energy source for the free-living, nitrogen-fixing bacteria in the rhizosphere. The concept of an aerobic-anaerobic interfacial area for active nitrogen fixation at the surface of flooded soil [217, 218] can apparently apply to the rice rhizosphere. The products of anaerobic respiration in flooded soil may diffuse to the aerobic zone near the root surface where the aerobic N_2-fixing bacteria may fix more nitrogen with a given amount of energy material than anaerobic N_2-fixing bacteria [218]. Apparently an anaerobic site in combination with an aerobic environment is essential for a significant rate of nitrogen fixation by bacteria [217-219]. More soluble products from the degradation of cellulose accumulate when oxygen is deficient [220] than under aerobic conditions. The free-living, N_2-fixing bacterium, *Azotobacter*, fixes the atmospheric nitrogen more efficiently when the oxygen concentration is lower in the soil than in the air [221]. Parker and Scutt [222] suggested that oxygen and nitrogen may compete as terminal hydrogen acceptors in the N_2-fixing bacterium. Considering the supply of molecular nitrogen and carbohydrates and the large area of the aerobic-anaerobic interface, nitrogen fixation by bacteria in the rice rhizosphere is perhaps biochemically and agronomically important.

Autotrophic nitrifying bacteria can function in the rice rhizosphere if enough molecular oxygen is available. Arima et al. [223] reported that the root tissues themselves, not rhizosphere bacteria, possessed an enzyme system that oxidized ammonium to nitrate. Although nitrate is not commonly used in rice culture, nitrate may be

present in the rhizosphere of rice. Nitrification in the rice rhizosphere awaits further investigation.

The roots of lowland plants that have developed an air-transporting system show much less nitrate reduction activity than roots
of upland plants because of the oxidizing activity [224]. Detaching
the tops or blocking the air-transporting systems with olive oil or
fluid paraffin increases the nitrate reduction activity [224]. These
data suggest that the depression of molecular oxygen is a major
factor increasing nitrate respiration in the rhizosphere. Arikado
[224] proposed that the root tissues themselves, as well as bacteria,
reduce nitrate for anaerobic respiration if they suffer from oxygen
deficiency, because both upland and lowland plants possess the
nitrate-reducing enzyme. Because nitrate reductase is an adaptive
enzyme formed in rice seedlings [225], the presence of nitrite reductase may indicate the presence of nitrate and possibly the existence of nitrification in the rice rhizosphere. Whether the nitrate
reduction results from the activity of rhizosphere bacteria or of
roots, or of both, cannot be ascertained unless experiments are
carried out under completely aseptic conditions.

The rhizosphere bacteria probably play a role in mineralizing
organic nitrogen in flooded soils, since rice roots have no enzymes
that break down amino acids [226]. The effect of rice roots on the
mineralization and immobilization of nitrogen in the rice rhizosphere
has not been studied. Neither the bacterial oxidation of such minerals as manganese, iron, sulfide, methane, and hydrogen nor the
presence of bacteria that oxidize these compounds in the rice rhizosphere has ever been reported. The heterotrophic characters of
strictly autotrophic bacteria that oxidize iron, sulfide, and molecular hydrogen have been reported [227-229]. These bacteria if
present, may also carry out biochemical functions in the rice rhizosphere. If ferrous iron is oxidized to ferric iron in the rhizosphere either by the roots or by bacteria, a decline in the oxidizing
activity of the rice rhizosphere would cause a direct flow of ferrous
iron into plant. The oxidizing activity of roots may correlate with
the susceptibility of rice plant to iron toxicity.

Rice plants greatly influence sulfur transformation in the rice
rhizosphere. Hydrogen sulfide, which is toxic to rice [230, 231] is
produced in flooded soils by microbial reduction of sulfate or from
sulfur-containing organic materials. Remarkably less sulfide is produced in the flooded soils with rice plants than in unplanted soils,
but once the rice rhizosphere becomes less oxidative more sulfide is
produced in planted soils than in unplanted soils [232]. The oxidizing
activity of the roots of nitrogen-deficient rice plants is low; accordingly, the sulfate-reducing activity in the rhizosphere increases
[232,233]. Higher root oxidizing activity decreases the toxic
effect of sulfide to the rice plant [233, 234]. Dommergues et al.
[235] reported that when some saline soils in Tunisia were flooded
H_2S production in the maize rhizosphere increased.

The populations of aerobic and anaerobic phosphate-dissolving
bacteria increase in the rice rhizosphere [236]. The activity of
bacteria that mineralize organic phosphates also increases, although
rice roots themselves possess phosphatase activities [226, 237].
The biochemical mechanism in the rhizosphere that makes inorganic
phosphate soluble is not fully understood.

C. Inorganic and Organic Toxins

The widely distributed nutritional disorders of the rice plant
in Asia frequently occur in rice fields that have high organic matter,
poor drainage systems, and highly reduced soil conditions [238, 239].
More intensive microbial functions of anaerobic respiration and fermentation probably take place under such soil conditions than in
normal flooded soils. After successive reduction of soil components,
the soils reach a reduced condition and produce various reduced inorganic compounds and anaerobic metabolites. The phytotoxic substances, both inorganic and organic, produced in reduced soil probably
affect the metabolic activity of rice roots and may interfere with
nutrient uptake.

Addition of organic materials to flooded soils increases the
concentration of soluble ferrous iron, manganous ions, and hydrogen

sulfide [240]; so rice plants take up more iron and manganese [241]. The excess ferrous iron produced in flooded soil is considered the cause of the iron toxicity of rice, called bronzing, in Ceylon [242]. Decrease of the root activity may aggravate iron toxicity in the soil because the rhizosphere of the rice plant becomes favorable for the anaerobic respiration of iron by iron-reducing bacteria. Sulfide formation in the rhizosphere can induce iron toxicity in rice plants by reducing the root-oxidizing activity [243-245].

Despite the active reduction of manganese in flooded soils, manganese toxicity to the rice plant is apparently less of a problem than iron toxicity in the rice culture. A high level of iron in flooded soils may counteract manganese uptake by the rice plant [239, 246]. A report, however, suggested that manganese toxicity may be the cause of a physiological disorder of rice plant in India [247]. Hydrogen sulfide can directly cause a physiological disorder in the rice plant in Akiochi soils in Japan [238]. The injury caused by hydrogen sulfide decreases the metabolic absorption by the rice plant of phosphate, potassium, and nitrogen, in that order [230, 231]. Large amounts of free hydrogen sulfide were found in flooded rice fields into which forage residues had been incorporated [248].

Tsubota [249] found that the reduced compounds from phosphate, such as phosphite, hypophosphite, and phosphine, injured the rice plant in a water culture experiment. Some problems in rice growth are apparently due to excess ammonia produced in soil [238]. Carbon dioxide, mostly produced as a result of microbial metabolism in soil, is highly soluble in water and is considered toxic to rice [250-252]. Carbonic acid retards the uptake of manganese and potassium [253]. Among the various gas components that evolve in flooded soil after flooding, methane is the only known gaseous hydrocarbon to accumulate in large amounts. Methane inhibits the growth of tomato [254] and barley, but not that of the rice plant [250]. Recently, Smith and Russell [255] found that ethylene accumulated in the soil under anaerobic conditions in amounts that could injure the roots of some plants.

Under flooded soil conditions, the organic constituents of
plants convert to a variety of intermediates in soil and in the
rhizosphere. Some of these reduced compounds may be toxic to rice
plants. The phytotoxic substances are produced in flooded soil most
abundantly in the early stages of decomposition of plant residues
[256, 257]. The soil metabolites most widely studied for their
phytotoxic effect on the growth of the rice plant are organic acids,
the lower volatile fatty acids. Such organic acids as acetic, propionic, and butyric acid should play a key role in the occurrence of
root damage in poorly drained sandy and peaty rice paddies [104, 258,
259]. These organic acids retard the uptake of nutrients by rice
plants [15, 253]. The toxic effects of organic acids depend on pH
level and the amount and kind of acids [104, 253]. Because the soil
pH of flooded soils generally becomes nearly neutral after flooding
[260] and because organic acids are more toxic at lower soil pH
values the harmful effect of organic acids in flooded soil may not
be a particularly important problem [253]. However, the injurious
effect on the rice plant of green manure application in flooded soils
is probably caused by organic acids produced during fermentation of
green manure [175, 248, 261]. The rice field into which forage residues are incorporated accumulates acetic acid and butyric acid in
amounts injurious to rice roots within a few weeks after flooding
[248]. If the organic matter is allowed to decompose for 50 days
before flooding, however, organic acids do not accumulate to toxic
levels [248]. The formation of alcohol [117] and of phenolic acids
[110] in flooded soil also affects rice plant growth. However, the
effect of the phytotoxic metabolites on the growth of the rice plant
and the rice yield in paddy fields awaits further study.

REFERENCES

The review of literature reported in this chapter was completed in 1971.

1. S. Ishizawa and H. Toyoda, *Bull. Natl. Inst. Agri. Sci. Ser. B., 14,* 203 (1964).
2. T. Suzuki, *Japan Agri. Res. Quarterly, 2,* 8 (1967).
3. C. Furusaka, T. Hattori, K. Sato, H. Yamagishi, R. Hattori, I. Nioh and M. Nishio, *Rep. Inst. Res. Tohoku Univ., 20,* 89 (1969).
4. P. D. Kumar and N. M. Bose, *Indian J. Agri. Sci., 8,* 487 (1938).
5. G. Rangaswami and R. Venkatesan, *Microorganisms in paddy soil,* Annamalai University, Annamalainagar, Madras, India, 1966.
6. S. M. Taha, S. A. Z. Mahmoud, and A. N. Ibrahim, *Plant Soil, 26,* 33 (1967).
7. Y. Takai, *J. Sci. Soil Manure Japan, 23,* 37 (1952).
8. Y. Takai, T. Koyama, and T. Kamura, *Agri. Chem. Soc. Japan, 29,* 967 (1955).
9. Y. Takai, T. Koyama, and T. Kamura, *Agri. Chem. Soc. Japan, 31,* 211 (1957).
10. Y. Takai and T. Kamura, *Kagaku (in Japanese), 31,* 618 (1961).
11. Y. Takai and T. Kamura, *Folia Microbiol., 11,* 304 (1966).
12. Y. Takai, *Agri. Tech. (in Japanese), 16,* 1, 51, 122, 162, 213, (1961).
13. W. J. Payne, *Ann. Rev. Microbiol., 24,* 17 (1970).
14. M. Shioiri, *J. Sci. Soil Manure Japan, 16,* 104 (1942).
15. S. Mitsui, *Inorganic Nutrition, Fertilization and Soil Amelioration for Lowland Rice,* 2nd ed., Yokendo Press, Tokyo, 1955.
16. W. H. Pearsall, *Empire Exp. Agri., 18,* 289 (1950).
17. H. Greene, *Nature (London), 186,* 511 (1960).
18. A. D. Scott and D. D. Evans, *Soil Sci. Soc. Amer. Proc., 19,* 7 (1955).
19. A. L. Chapman and D. S. Mikkelsen, *Crop Sci., 3,* 392 (1963).
20. W. H. Patrick, Jr., and M. B. Sturgis, *Soil Sci. Soc. Amer. Proc., 19,* 59 (1955).
21. D. R. Bouldin, *J. Ecol., 56,* 77 (1968).
22. J. W. T. Wimpenny, in *Microbial Growth,* 19th Symposium of the Society for General Microbiology, (P. Meadow and S. J. Pirt, Eds.), Vol. 161, Cambridge Univ. Press, Cambridge, 1969.
23. W. Armstrong, *Physiol. Plantarum, 22,* 296 (1969).
24. W. Armstrong, *Nature (London), 204,* 801 (1964).
25. H. J. Dittmer, *Amer. J. Bot., 25,* 654 (1938).
26. H. W. Doelle, *Bacterial Metabolism,* Academic, New York, 1969.
27. D. S. Mikkelsen and D. S. Finfrock, *Agron. J., 49,* 296 (1957).
28. G. V. Simsiman, S. K. De Datta, and J. C. Moomaw, *J. Agri. Sci. (Camb.), 69,* 189 (1967).
29. S. Patnaik, *Proc. Indian Acad. Sci. Sec. B., 61,* 25 (1965).
30. D. A. Rennie and Mr. Fried, Int. Symp. Soil Fert. Eval. Proc., *New Delhi, India, 1,* 639 (1971).

31. C. T. Abichandani and S. Patnaik, *J. Ind. Soc. Sci.*, *6*, 87
     (1958).
32. W. H. Patrick, Jr. and I. C. Mahapatra, *Advan. Agron.*, *20*,
     323 (1968).
33. F. E. Broadbent and D. S. Mikkelsen, *Agron. J.*, *60*, 674 (1968).
34. A. H. Manzano, Ph.D. Thesis, University of California, 1968.
35. M. E. Tusneem, Ph.D. Thesis, Louisiana State University, Baton
     Rouge, 1970.
36. H. Broeshart, *Nitrogen-15 in Soil and Plant Studies*, p. 47, IAEA
     Vienna, 1971.
37. F. E. Broadbent and M. E. Tusneem, *Soil Sci. Soc. Amer. Proc.*,
     35, 922 (1971).
38. S. Motomura, *J. Sci. Soil Manure Japan*, *35*, 431 (1964).
39. S. Motomura, *J. Sci. Soil Manure Japan*, *37*, 263 (1966).
40. R. S. Wolfe, in *Principles and Application in Aquatic Micro-
     biology* (H. H. Heukelekian and N. S. Dondero, Eds.), p. 82,
     Wiley, New York, 1963.
41. M. Alexander, *Introduction to Soil Microbiology*, Wiley, New
     York, 1961.
42. D. Wertlieb and W. Vishniac, *J. Bact.*, *93*, 1722 (1967).
43. F. R. Moorman, *Soil Sci.*, *95*, 271 (1963).
44. S. Vangnai, M.S. Thesis, College of Agriculture, University of
     the Philippines, Los Baños, Laguna, 1966.
45. M. G. R. Hart, *Plant Soil*, *11*, 215 (1959).
46. C. Bloomfield, *J. Soil Sci.*, *20*, 207 (1969).
47. W. E. Connell and W. H. Patrick, Jr., *Soil Sci. Soc. Amer.
     Proc.*, *33*, 711 (1969).
48. K. Y. Chen and J. C. Morris, *5th Int. Water Poll. Res. Conf.
     Proc. (San Francisco) III*, *32*, 1 (1970).
49. V. B. D. Skerman, J. Lack, and N. Millis, *Austral. J. Biol.
     Res.*, *4*, 511 (1951).
50. C. M. Gilmour, R. P. Bhatt, and J. V. Mayeux, *Nature (London)*,
     203, 55 (1964).
51. F. E. Broadbent and F. E. Clark, in *Soil Nitrogen* (W. V. Bar-
     tholomew and F. E. Clark, Eds.), No. 10, p. 344, Amer.
     Soc. Agron., 1965.
52. I. C. MacRae, R. R. Ancajas, and S. Salandanan, *Soil Sci.*,
     105, 327 (1968).
53. T. Nagata and K. Matsuda, *J. Sci. Soil Manure Japan*, *26*, 121
     (1955).
54. A. J. Kluyver and W. Verhoeven, *Antonie van Leeuwenhoek*, *20*,
     241 (1954).
55. P. W. Arnold, *J. Soil Sci.*, *5*, 116 (1954).
56. J. Wijler and C. C. Delwiche, *Plant Soil*, *5*, 155 (1954).
57. C. C. Delwiche, *J. Bact.*, *77*, 55 (1959).
58. A. S. Corbet, *Biochem. J.*, *29*, 1086 (1935).
59. T. Yoshida and M. Alexander, *Soil Sci. Soc. Amer. Proc.*, *34*,
     880 (1971).
60. J. H. Jordan, Jr., W. H. Patrick, Jr., and W. H. Willis, *Soil
     Sci.*, *104*, 129 (1967).
61. P. J. G. Mann and J. H. Quastel, *Nature (London)*, *158*, 154 (1946).

62. R. M. Hochster and J. H. Quastel, *Arch. Biochem. Biophys.*, *36*, 132 (1952).
63. M. P. Silverman and H. L. Ehrlich, *Advan. Appl. Microbiol.*, *6*, 153 (1964).
64. R. B. Trimble and H. L. Ehrlich, *Appl. Microbiol.*, *16*, 695 (1968).
65. L. N. Mandal, *Soil Sci.*, *91*, 121 (1961).
66. S. Gotoh and K. Yamashita, *Soil Sci. Plant Nutr. Japan*, *12*, 230 (1966).
67. S. M. Bromfield, *J. Soil Sci.*, *5*, 129 (1954).
68. K. Yamanaka and S. Motomura, *J. Sci. Soil Manure Japan*, *29*, 104 (1958).
69. T. Kamura and T. Takai, *J. Sci. Soil Manure Japan*, *31*, 499 (1960).
70. S. Motomura, *Bull. Natl. Insti.*, *Agri. Sci.*, *Ser. B*, *21*, 1 (1969).
71. T. Asami and Y. Takai, *J. Sci. Soil Manure Japan*, *41*, 48 (1970).
72. I. Yamane, *Rep. Inst. Agri. Res. Tohoku Univ.*, *21*, 39 (1970).
73. F. N. Ponnamperuma, E. Martinez, and T. Loy, *Soil Sci.*, *101*, 421 (1965).
74. T. Asami, *J. Sci. Soil Manure Japan*, *41*, 7 (1970).
75. L. E. Allison and G. D. Scarseth, *J. Amer. Soc. Agron.*, *34*, 616 (1942).
76. C. Bloomfield, *J. Soil Sci.*, *2*, 196 (1951).
77. J. C. G. Ottow, *Oecologia (Berl)*, *6*, 164 (1971).
78. J. C. G. Ottow, *Z. Allg. Microbiol.*, *8*, 441 (1968).
79. J. C. G. Ottow, *Die Naturwiss*, *7*, 371 (1969).
80. J. C. G. Ottow, *Z. Pflernähr. Düng. Bodenk.*, *124*, 238 (1969).
81. A. F. DeCastro and H. L. Ehrlich, *Antonie van Leeuwenhoek*, *36*, 317 (1970).
82. W. E. Connell and W. H. Patrick, Jr., *Science*, *159*, 86 (1968).
83. C. Furusaka, *Bull. Inst. Agri. Res. Tohoku Univ.*, *19*, 101 (1968).
84. J. R. Postgate, *Ann. Rev. Microbiol.*, *13*, 505 (1959).
85. J. R. Postgate, *Bact. Rev.*, *29*, 425 (1965).
86. H. D. Peck, Jr., *Bact. Rev.*, *26*, 67 (1962).
87. H. Shiga and S. Suzuki, *Bull. Chuogoku Agri. Exp. Sta.*, *Ser. A*, *9*, 185 (1963).
88. I. Yamane and I. Sato, *Sci. Rep. Inst. Tohoku Univ.*, *Ser. D*, *12*, 73 (1961).
89. Y. Takai, *Soil Sci. and Plant Nutr.*, *16*, 238 (1970).
90. G. Tsubota, *Bull. Toichigi Agri. Exp. Sta.*, *2*, 1 (1958).
91. K. Tensho and K. L. Yeh, *Soil Sci. Plant Nutr. Tokyo*, *16*, 30 (1970).
92. K. Tensho and K. L. Yeh, *Radioisotopes*, *19*, 574 (1970).
93. K. Tonomura, T. Nakagami, F. Futai, and K. Maeda, *J. Ferment. Technol.*, *46*, 506 (1968).
94. K. Tonomura and F. Kanzaki, *Biochem. Biophys. Acta*, *184*, 227 (1969).
95. C. E. Woolfolk and H. R. Whiteley, *J. Bact.*, *84*, 647 (1962).
96. W. A. Wood, in *The Bacteria* (I. C. Gunsalus and R. Y. Stanier, Eds.), Vol. 2, p. 59, Academic, New York, 1961.
97. R. Y. Stanier, M. Doudoroff, and E. A. Adelberg, *The Microbial World*, 2nd ed., Prentice-Hall, Englewood Cliffs, New Jersey, 1965.

98. H. A. Barker, Bacterial Fermentations, CIBA, Lectures in Micro-
    bial Biochemistry, Wiley, New York, 1966.
99. B. Wada and T. Iida, *J. Sci. Soil Manure Japan*, *7*, 1 (1933).
100. S. Mitsui, K. Kumazawa, and T. Ishihara, *J. Sci. Soil Manure
     Japan*, *24*, 45 (1953).
101. S. Mitsui, K. Kumazawa, and M. Mukai, *J. Sci. Soil Manure Japan*,
     30, 345 (1959).
102. S. Mitsui, K. Kumazawa, and T. Ishida, *J. Sci. Soil Manure
     Japan*, *30*, 411 (1959).
103. Y. Takijima, *J. Sci. Soil Manure Japan*, *31*, 435 (1960).
104. Y. Takijima, M. Shiojima, and Y. Arita, *J. Sci. Soil Manure
     Japan*, *31*, 441 (1960).
105. Y. Takijima, *Bull. Natl. Agri. Sci. Japan, Ser. B*, *13*, 117
     (1963).
106. S. Motomura, Y. Akiyama, and K. Yamanaka, *J. Sci. Soil Manure
     Japan*, *32*, 605 (1961).
107. K. Sato and I. Yamane, *J. Sci. Soil Manure Japan*, *26*, 509 (1956).
108. S. Goto and Y. Onikura, *Bull. Kyushu Agri. Exp. Sta. Japan*, *13*,
     173 (1967).
109. J. P. Hollis and R. Rodriguez-Kabana, *Plant Pathol.*, *57*, 841
     (1967).
110. T. S. C. Wang, S. Y. Cheng, and H. Tung, *Soil Sci.*, *104*, 138
     (1967).
111. K. Fujii, M. Kobayashi, M. Z. Hague, and E. Takahashi, *J. Sci.
     Soil Manure Japan*, *41*, 286 (1970).
112. F. J. Stevenson, in *Soil Biochemistry* (A. D. McLaren and G. H.
     Peterson, Eds.), p. 119, Dekker, New York, 1967.
113. A. H. Rose, *Chemical Microbiology*, Butterworths, London, 1965.
114. Y. Takai, J. Macura, and F. Kung, *Folia Microbiologia*, *14*, 327
     (1969).
115. D. J. Greenwood and H. Lees, *Plant Soil*, *12*, 69 (1961).
116. H. A. Barker, *J. Biol. Chem.*, *137*, 153 (1941).
117. T. S. C. Wang and T. T. Chuang, *Soil Sci.*, *104*, 40 (1967).
118. H. A. Barker, in *The Bacteria* (I. C. Gunsalus and R. Y. Stanier,
     Eds.), Vol. 2, p. 151, Academic, New York, 1961.
119. M. Alexander, *Appl. Microbiol.*, *7*, 35 (1965).
120. D. T. Gibson, *Science*, *161*, 1093 (1968).
121. F. M. Clark and L. R. Fina, *Arch. Biochem. Biophys.*, *36*, 26
     (1952).
122. L. R. Fina and A. M. Fiskins, *Arch. Biochem. Biophys.*, *91*,
     163 (1960).
123. P. M. Nottingham and R. E. Hungate, *J. Bacteriol.*, *98*, 1170
     (1969).
124. B. F. Taylor, W. L. Campbell, and I. Chinoy, *J. Bact.*, *102*,
     430 (1970).
125. P. L. Dutton and W. C. Evans, *Biochem. J.*, *113*, 525 (1969).
126. M. Guyer and G. D. Hegeman, *J. Bacteriol.*, *99*, 906 (1969).
127. G. D. Hegeman, *Arch. Mikrobiol.*, *59*, 143 (1967).
128. J. Chmielowski and W. Wasilewski, *Zesz. Nauk. Politech. Slaska.
     Inz. Sanit. (Polon.).*, *9*, 95 (1966), From *Chem. Abst.*, *67*,
     93807q (1967).

129. C. S. Boruff and A. M. Buswell, *J. Amer. Chem. Soc.*, *56*, 886 (1934).
130. M. Levine, G. H. Nelson, D. Q. Anderson, and P. B. Jacob, *Ind. Eng. Chem.*, *27*, 195 (1935).
131. W. H. Harrison and P. A. S. Aiyer, *Mem. Dept. Agri. India Chem.*, *Ser. 3*, p. 65 (1913).
132. I. Yamane and K. Sato, *J. Sci. Soil Manure Japan*, *32*, 264 (1961).
133. K. Sato and I. Yamane, *J. Sci. Soil Manure Japan*, *37*, 547 (1966).
134. H. A. Barker, *Bacterial Fermentations*, John Wiley, New York, 1956.
135. P. L. McCarty, in *Principles and Applications in Aquatic Microbiology* (H. Heukelekian and N. C. Dondero, Eds.), p. 314, Wiley, New York, 1964.
136. T. C. Stadtman, *Ann. Rev. Microbiol.*, *21*, 121 (1967).
137. A. M. Buswell and F. W. Sollo, *J. Amer. Chem. Soc.*, *70*, 1778 (1949).
138. T. C. Stadtman and H. A. Barker, *J. Bact.*, *61*, 81 (1951).
139. T. C. Stadtman and H. A. Barker, *Arch. Biochem.*, *21*, 256 (1949).
140. M. J. Pine and H. A. Barker, *J. Bacteriol.*, *71*, 644 (1956).
141. I. Yamane and K. Sato, *Soil Sci. Plant Nutr. Japan*, *9*, 32 (1963).
142. J. T. Moraghan, *Plant Soil*, *31*, 1 (1969).
143. R. G. Bell, *Soil Biol. Biochem.*, *1*, 105 (1969).
144. D. Laskowski and J. T. Moraghan, *Plant Soil*, *27*, 357, (1967).
145. C. A. Edwards, *Residue Rev.*, *13*, 83 (1966).
146. E. H. Marth, *Residue Rev.*, *9*, 89 (1965).
147. J. G. Saha, *Residue Rev.*, *26*, 89 (1969).
148. K. Raghu and I. C. MacRae, *Science*, *154*, 263 (1966).
149. I. C. MacRae, K. Raghu, and T. F. Castro, *J. Agri. Food Chem.*, *15*, 911 (1967).
150. T. Yoshida and T. F. Castro, *Soil Sci. Soc. Amer. Proc.*, *34*, 440 (1970).
151. I. C. MacRae, K. Raghu, and E. M. Bautista, *Nature (London)*, *221*, 859 (1969).
152. N. Sethunathan, E. M. Bautista, and T. Yoshida, *Can. J. Microbiol.*, *15*, 1349 (1969).
153. F. J. Oppenworth, *Nature (London)*, *173*, 1001 (1954).
154. J. Sternberg and C. W. Kearns, *J. Econ. Entomol.*, *49*, 548 (1956).
155. R. G. Bridges, *Nature (London)*, *184*, 1337 (1959).
156. P. L. Grover and P. Sims, *Biochem J.*, *96*, 521 (1965).
157. W. N. Yule, M. Chiba, and H. B. Morley, *J. Agri. Food Chem.*, *15*, 1000 (1967).
158. C. I. Chacko, J. L. Lockwood, and M. Zabik, *Science*, *154*, 893 (1966).
159. J. L. Mendel and M. S. Walton, *Science*, *151*, 1527 (1966).
160. B. T. Johnson, R. N. Goodman, and H. S. Goldberg, *Science*, *157*, 560 (1967).
161. J. R. Plimmer, P. C. Kearney, and D. W. van Endt, *J. Agri. Food Chem.*, *16*, 594 (1968).
162. R. C. Braunberg and V. Beck, *J. Agri. Food Chem.*, *16*, 451 (1968).
163. D. W. Hill and P. L. McCarty, *J. Water Poll. Cont. Fed.*, *39*, 1259 (1967).

164. L. W. Newland, G. Chesters, and G. B. Lee, *J. Water Poll. Cont. Fed.*, *41*, R174 (1969).
165. I. C. MacRae, K. Raghu, and T. F. Castro, *J. Agri. Food Chem.*, *15*, 911 (1967).
166. T. F. Castro and T. Yoshida, *J. Agri. Food Chem.*, *19*, 1168 (1971).
167. P. R. Datta, E. P. Lang, and A. K. Klein, *Science*, *145*, 1052 (1964).
168. B. J. Kallman and A. K. Andrews, *Science*, *141*, 1050 (1963).
169. J. H. V. Stenersen, *Nature (London)*, *207*, 660 (1965).
170. G. Wedemeyer, *Science*, *152*, 647 (1966).
171. G. Wedemeyer, *Appl. Microbiol.*, *15*, 569 (1967).
172. W. H. Ko and J. L. Lockwood, *Can. J. Microbiol.*, *14*, 1069 (1968).
173. W. D. Guenzi and W. E. Beard, *Science*, *156*, 1116 (1967).
174. J. F. Parr, G. H. Willis, and S. Smith, *Soil Sci.*, *110*, 306 (1970).
175. The International Rice Research Institute, *Annual Report*, p. 47, Los Baños, Laguna, Philippines, 1970.
176. R. P. Miskus, D. P. Blair, and J. E. Casida, *J. Agri. Food Chem.*, *13*, 481 (1965).
177. J. L. Mendel, A. K. Klein, J. T. Chen, and M. S. Walton, *J.A.O. A.C.*, *50*, 897 (1967).
178. F. Matsumura and G. M. Boush, *Science*, *156*, 959 (1967).
179. F. Matsumura, G. M. Boush, and A. Tai, *Nature (London)*, *219*, 965 (1968).
180. J. R. W. Miles, C. M. Tu, and C. R. Harris, *J. Econ. Entomol.*, *62*, 1334 (1969).
181. G. Wedemeyer, *Appl. Microbiol.*, *16*, 661 (1967).
182. M. Alexander, *Soil Biology*, p. 209, UNESCO, 1969.
183. L. W. Getzin, *J. Ecol. Entomol.*, *60*, 505 (1967).
184. N. Sethunathan and T. Yoshida, *J. Agri. Food Chem.*, *17*, 1192 (1969).
185. D. A. Graetz, G. Chesters, T. C. Daniel, L. W. Newland, and G. B. Lee, *J. Water Poll. Conf. Res.*, *42*, R76 (1970).
186. M. K. Arhmed, J. E. Casida, and R. E. Nicholas, *J. Agri. Food Chem.*, *6*, 740 (1958).
187. E. P. Lichenstein and K. R. Schulz, *J. Econ. Entomol.*, *57*, 618 (1964).
188. D. L. Mick and P. A. Dahm, *J. Econ. Entomol.*, *63*, 1155 (1970).
189. B. M. Zuckerman, K. Deubert, M. Mackiewicz, and H. Gunner, *Plant Soil*, *33*, 273 (1970).
190. The International Rice Research Institute, *Annual Report*, p. 23, Los Baños, Laguna, Philippines, 1971.
191. F. E. Clark, *Advan. Agron.*, *1*, 241 (1949).
192. M. I. Timonin, in *Microbiology and Soil Fertility* (C. M. Gilmour and O. N. Allen, Eds.), p. 135, Oregon State University Press, 1964.
193. Z. Tesic, in *Plant Microbes Relationships* (J. Macura and V. Vancura, Eds.), p. 15, Publishing House of Czechoslovakia Acad. Sci., Prague, 1965.

194. A. D. Rovira and B. M. McDougall, in *Soil Biochemistry* (A. D. McLaren and G. H. Peterson, Eds.), p. 417, Dekker, New York, 1967.
195. T. Yamasaki, *Bull. Natl., Inst. Agri. Sci., Ser. B, No. 1*, (1952).
196. M. H. Van Raalte, *Ann. Bot. Gardens, Buitenzorg, 51*, 43 (1941).
197. R. Aimi, *Crop Sci. Soc. Japan Proc., 29*, 51 (1960).
198. H. Arikado, *Bull. Fac. Agri., Mie Univ. Japan, 24*, 17 (1961).
199. C. R. Jensen, L. H. Stolzy, and J. Jetey, *Soil Sci., 103*, 23 (1967).
200. D. A. Barber, M. Ebert, and N. T. S. Evans, *J. Exp. Bot., 13*, 397 (1962).
201. K. Arashi and H. Nitta, *Crop Sci. Soc. Japan Proc., 24*, 78 (1955).
202. H. Arikado, *Bull. Fac. Agri., Mie Univ. Japan, 19*, 1 (1959).
203. D. R. Bouldin, *F.A.O./I.A.E.A. Tech. Rep. Ser., 65*, 128 (1958).
204. S. Mitsui and K. Tensho, *J. Sci. Soil Manure Japan, 22*, 301 (1952).
205. G. Izawa, Y. Oji, and S. Okamoto, *J. Sci. Soil Manure Japan, 37*, 552 (1966).
206. K. Sato, *Crop Sci. Soc. Japan Proc., 21*, 16 (1952).
207. K. Inada, *Bull. Natl. Inst. Agri. Sci., Ser. D, 16*, 19 (1967).
208. A. D. Rovira, *Soils Fert., 25*, 167 (1962).
209. R. Andal, K. Bhuvaneswari, and N. S. Subba-Rao, *Nature (London), 178*, 1063 (1956).
210. I. C. MacRae and T. F. Castro, *Plant Soil, 26*, 317 (1967).
211. Th. Alberda, *Plant Soil, 5*, 1 (1953).
212. S. P. Chakraborty and S. P. Sen Guppa, *Nature (London), 184*, 2033 (1959).
213. M. G. Hart and E. H. Robert, *Nature (London), 189*, 598 (1961).
214. T. Yoshida and R. R. Ancajas, *Soil Sci. Soc. Amer. Proc., 35*, 156 (1971).
215. Y. Dommergues, J. Balandreau, G. Rinaudo, and P. Weinhard, *Twelfth Pacific Science Congress, Canberra, Australia, Record of Proceedings, Abstracts of Papers, 1*, p. 46, 1971.
216. J. Döbereiner and A. P. Ruschel, *Rev. Brasil. Biol., 21*, 397 (1961).
217. W. A. Rice, E. A. Paul, and L. R. Wetter, *Can. J. Microbiol., 13*, 829 (1967).
218. F. R. Magdoff and D. R. Bouldin, *Plant Soil, 33*, 49 (1970).
219. N. J. Barrow and D. S. Jenkinson, *Plant Soil, 16*, 258 (1962).
220. H. L. Jensen and R. J. Swaby, *Proc. Linnean Soc. N. S. Wales, 66*, 89 (1941).
221. C. A. Parker, *Nature (London), 173*, 780 (1954).
222. C. A. Parker and P. B. Scutt, *Biochem. Biophys. Acta, 38*, 230 (1960).
223. Y. Arima, K. Kumazawa, and S. Mitsui, *Ann. Meeting, Soc. Sci. Manure Japan, Abst. No. 15*, p. 70 (1969).
224. H. Arikado, *Bull. Fac. Agri. Mie Univ., 21*, 1 (1960).
225. P. S. Tang and H. Y. Wu, *Nature (London), 179*, 1355 (1957).
226. A. Okuda, M. Yamaguchi, and S. Yong-Gil, *J. Sci. Soil Manure Japan, 35*, 311 (1964).

227. R. M. Borichewski and W. W. Umbreit, *Arch. Bioch. Biophys, 116,* 97 (1966).

228. F. Shafia and R. F. Wilkinson, Jr., *J. Bact., 97,* 256 (1968).

229. H. H. Ramsey, *Antonie van Leeuwenhoek, 34,* 71 (1968).

230. S. Mitsui, S. Aso, and K. Kumazawa, *J. Sci. Soil Manure Japan, 22,* 46 (1951).

231. H. Okajima and S. Takagi, *Bull. Inst. Agri. Tohoku Univ., 5,* 139 (1953).

232. H. Shiga and S. Suzuki, *Bull. Chugoku Agri. Exp. Sta., Ser. A, 9,* 185 (1963).

233. H. Okajima, *J. Sci. Soil Manure Japan, 29,* 175 (1958).

234. I. Baba, *Crop Sci. Soc. Japan Proc., 23,* 167 (1955).

235. Y. Dommergues, R. Combremont, G. Beck, and C. Ollat, *Rev. Ecol. Biol. Sol., 6,* 115 (1969).

236. K. Raghu and I. C. MacRae, *J. Appl. Microbiol., 29,* 582 (1966).

237. T. Hayashi and Y. Takijima, *J. Sci. Soil Manure Japan, 27,* 15 (1956).

238. I. Baba, K. Inada, and K. Tajima, *Mineral Nutrition of the Rice Plant,* p. 173, John Hopkins Univ. Press, Baltimore, 1964.

239. A. Tanaka and S. Yoshida, *International Rice Research Institute Tech. Bull. No. 10,* Los Baños, Laguna, Philippines, 1970.

240. L. N. Mandal, *Soil Sci., 91,* 121 (1961).

241. F. E. Clark, D. C. Nearpas, and A. W. Specht, *Agron J., 49,* 586 (1957).

242. F. N. Ponnamperuma, R. Bradfield, and M. Peech, *Nature (London), 175,* 265 (1955).

243. A. Tanaka, R. P. Mulleriyawa, and T. Yasu, *Soil Sci. Plant Nutr. Japan, 14,* 1 (1968).

244. Y. D. Park and A. Tanaka, *Soil Sci. Plant Nutr. Japan, 14,* 27 (1968).

245. K. Inada, *Crop Sci. Soc. Japan Proc., 33,* 315 (1965).

246. A. Tanaka and S. A. Navasero, *Soil Sci. Plant Nutr. Japan, 12,* 197 (1966).

247. S. Perumal, *Soil Sci., 91,* 218 (1961).

248. I. Yamane and K. Sato, *Rep. Inst. Agri. Res. Tohoku Univ., 21,* 79 (1970).

249. G. Tsubota, *Bull. Tochigi Agri. Exp. Sta., 5,* 65 (1961).

250. J. Vlamis and A. R. Davis, *Plant Physiol., 19,* 33 (1944).

251. H. T. Chang and W. E. Loomis, *Plant Physiol., 20,* 221 (1945).

252. B. Saito, *Bull. Kyushu Agri. Exp. Sta., 2,* 283 (1954).

253. A. Tanaka and S. A. Navasero, *Soil Sci. Plant Nutr. Japan, 13,* 25 (1967).

254. F. R. Forsyth and C. A. Evans, *Physiol. Plantarum, 22,* 1055 (1969).

255. K. A. Smith and R. S. Russell, *Nature (London), 222,* 769 (1969).

256. P. J. Welbank, *Weed Res., 3,* 205 (1963).

257. Z. A. Patrick, *Soil Sci., 111,* 13 (1971).

258. Y. Takijima, *Bull. Natl. Inst. Agri. Sci., Ser. B, 13,* 117 (1963).

259. Y. Takijima, *Soil Sci. Plant Nutr. Japan, 10,* 204 (1964).

260. F. N. Ponnamperuma, *Mineral Nutrition of the Rice Plant,* p. 295, John Hopkins Univ. Press, Baltimore, 1964.

261. A. D. Desai and T. S. Rao, *J. Indian Soc. Soil Sci., 5,* 147 (1957).

CHAPTER 4

BIOCHEMISTRY AND MICROBIOLOGY OF PEATS

P. H. Given   and   C. H. Dickinson

Fuel Science Section   Department of Botany
College of   University of Newcastle-upon-Tyne
Earth and Mineral Sciences   England
The Pennsylvania State University
University Park, Pennsylvania

I. INTRODUCTION: NATURE AND DISTRIBUTION OF PEATS

The soil scientist usually regards peat as a special kind of soil. It is a growth medium for higher plants that is rich in organic matter, usually waterlogged, and sometimes of low pH. Like any soil, its organic matter is derived primarily from plant debris and, pre-sumably, from microorganisms and animals. There are, however, enough essential differences to make it worthwhile to consider peats *sui generis*, a distinct phenomenon in their own right. Perhaps the most important difference is that in a peat-forming environment organic matter is accumulating: peats represent a break in the carbon cycle. Fairly well-preserved fragments of plant tissue are recognizable, either macroscopically or microscopically [1, 2], and relatively few organic particles appear to be colonized by actively growing microorganisms [3, 4].

It is clear from geological and paleobotanical studies that coals represent the effect of heat and pressure on peats when they become deeply buried under a succession of other sediments. If all coal formations originated as peat swamps, then peat accumulation has been a widespread and important phenomenon for perhaps 300 of the 400 million years since land plants first occurred. The total world reserves of coal now known, in seams greater than 30 cm in thickness and less than 1200 m below the earth's surface, and from all geological eras, are about 6×10^{12} metric tons, whereas the current annual land plant biomass is estimated to be about 4.5×10^{10} tons. Hence only a very small fraction of the total plant production has escaped the degradation process of the normal carbon cycle in this particular way.

In addition to understanding peat formation from a geological, botanical, and hydrological viewpoint, we should understand the bio-chemistry of peat: the components present, the alterations to plant debris as it becomes incorporated in peat, the extent of subsequent alterations in the peat profile, and the agents of these changes.

Peats are known in many types of environments and are derived

from many different types of plant associations. Although a class
of peat-forming plants cannot be defined, plants that contribute to
peat accumulations have in common that they tolerate waterlogged
habitats. If peat is to accumulate in substantial thickness over
perhaps several thousand years, the water table and the peat surface
must keep an approximately constant relationship. Either the underlying rock must be subsiding or the water table must be rising, because of a rise in sea level, for example. Peat can also accumulate
in mountain basins formed by other geomorphic processes, in valleys,
in flat-lying coastland, and in river deltas.

The succession of peat-forming plants that are found at any
site, given equivalent geologic and hydrologic conditions, will depend on what climatic zone the site is in. There are large peat
deposits in such tropical and sub-tropical areas as the Mississippi
delta, the Everglades of southern Florida, parts of central America,
Africa, along coasts bordering the Indian Ocean, Malaya, Borneo, and
Indonesia. These include the extensive marine mangrove and other
swamps; in contradistinction, the peat deposits in temperate latitudes
are rarely saline.

According to Nikonov [5] peats occupy about 180×10^6 hectares
or 0.7% of the land surface of the earth. Table 1 shows the distribution of peatlands in a large number of geographical areas, the
total being less than Nikonov's figure. In such regions as the
northern margin of the USSR, parts of Canada, and Minnesota in the
United States, glaciation has ground "kettles" out of the rock, and
when the glaciers receded, these filled with water. The new lakes
in time underwent eutrophication, so that various communities of
higher plants were successively able to inhabit the site, often
forming a succession of peat types. There would be a somewhat different characteristic relationship between water table and peat
sediment surface for each successive community.

There are peat deposits in the Caucasus, in southern USSR, but
it is not clear what proportion of the whole USSR peatlands these
represent: certainly quite a small proportion. Of the 5.2×10^6

TABLE 1

Distribution of Peat Deposits and the Extent of Peat
Occurrence on the Earth's Surface[a]

Country	Area of peat deposits, (million hectares)	Extent of peat occurrence in percentage of total area
USSR	71.5	3.2
Canada	10.0	1.0
Finland	7.0	19.0
Sweden	5.0	7.5
United States	5.2	0.7
Great Britain	2.4	10.0
Poland	1.5	4.7
Ireland	1.2	17.0
W. Germany	1.2	4.8
Indonesia	1.0	0.5
Congo	1.0	0.4
Norway	0.9	2.7
East Germany	0.5	4.5
Kenya, Uganda	0.5	0.7
Iceland	0.3	3.0
Japan	0.3	0.7
Pakistan	0.3	0.2
New Zealand	0.2	0.6
Denmark	0.1	4.6
Cuba	0.1	1.3
India	0.1	0.1
Italy	0.1	0.3
France	0.1	0.2
Brazil	0.1	0.1
China	0.1	0.1
	110.7	1.5

[a]Data from Ref. [6].

hectares shown for the United States, perhaps 1.5-2×10^6 are in subtropical climatic zones. Accepting this latter figure, it appears that a total of about 4.5-5×10^6 hectares of peat occur in tropical and subtropical areas; if one adds to this some contribution from southern USSR, peats in these climatic zones represent 5-8% of the total world peat resources.

It was, until comparatively recently, questioned by some whether peat could ever accumulate in moist tropical environments, on the grounds that rates of decomposition would be so high; it was not appreciated that productivity, particularly in estuarine and deltaic environments, can also be very high. In fact, through most of the earth's history, tropical and subtropical climatic zones have extended much further from the equator than they do now. Most of the major coal deposits of the northern hemisphere, formed in the Carboniferous, the Cretaceous, and the Tertiary, were laid down as peat deposits (often saline) in subtropical and tropical conditions. Therefore, peats in the warmer climatic zones have evidently represented a much larger proportion of total peats than is now the case. To the geochemist, who takes the whole of geological time as his province, present-day tropical and subtropical peats are considerably more significant than their relative abundance would suggest.

Unfortunately, the data necessary for forming a *Weltanschauung* (or world view) of peat formation do not exist. A search through references to biochemical and microbiological papers on peat in "Chemical Abstracts" for 1962-1972 reveals that a large proportion (80-90%?) are by authors working in the USSR or Poland. Only four papers were found (4%) dealing with peats in subtropical or tropical climatic zones (Florida, 3; Malaya, 1). The literature is thus overwhelmingly concerned with peats from north temperate zones, and less than justice can be done in this review to the substantial tropical peat deposits. The authors and their colleagues have made extensive studies of peat accumulations in the Florida Everglades. These will be relied on for such discussion of subtropical peats as is possible at this time.

II. HYDROLOGICAL AND BOTANICAL ASPECTS OF PEAT SYSTEMS: CLASSIFICATION

A. Classification of Peat Bogs in Temperate Zones

In the early part of this century, Weber [7] propounded the
theory that peat systems are related through their formation as part
of a regulated biotic process. This theory envisaged peat as formed
initially in open water leading to *niedermoore* (valley bog), which
grows through further peat accumulation to give *übergangsmoore* (transition bog) and eventually *hochmoore* (raised bog). Topography, together with water source, water table, nutrient supply and the life
forms of the plants themselves, has been used by Ratcliffe [8] to
delimit three similar groups of peat types. Fens are formed under
topogenous conditions, mires are formed under soligenous conditions
where there is a flow of groundwater, and bogs are characteristic of
ombrogenous situations, in regions of high rainfall, where they may
develop either on gently contoured uplands (blanket bog) or above
topogenous formations (raised bog).

Peats that developed as a result of a widespread climatic factor,
perhaps through microbial activity, were termed "climatic" or "zonal"
peats by Fraser [9], whereas "intrazonal" peats were considered to
be caused by the operation of local factors such as impeded drainage.
Fraser also described peat profiles in recognized pedalogical terms
using such factors as texture, structure, origin, degree of decomposition, and level of the water table to delimit each horizon.

High concentrations of calcium, magnesium and bicarbonate are
characteristically found in fen peat waters, whereas moss or raised
bog waters contain relatively low levels of these ions [10, 11]. Exchangeable nutrient concentrations have also been used to support
the divisions of peats into minerotrophic types, which include fens
and mires, and an ombrotrophic group of raised and blanket bogs.
The former usually have an exchangeable calcium-to-magnesium ratio
greater than 1.0, whereas in the latter it is less than 1.0 [12].
Volarovich et al. [13] give average contents of 25.4, 94.8, and
161.7 mEq/100 g dry peat for total Ca^{2+} plus Mg^{2+} in highland,

intermediate, and lowland peats, respectively.

The formation of peat in flowing groundwaters rich in dissolved plant nutrients leads to a rheophilous habitat, in transition habitats the supply of groundwater limits productivity, and ombrophilous formations depend on nutrient-poor rainwater [14]. Bellamy [15] accepts the same basic divisions and terms and he also stressed the relative concentrations of the dominant ions. He found that western European rheophilous mires were dominated by bicarbonate and calcium ions and that they could be divided into four subgroups depending upon the constancy (or otherwise) of the flow of groundwater and the usual level it attained in relation to the peat surface. Transition mires were characterized by relatively high concentrations of both calcium and sulfate ions whereas ombrophilous mires lack supplies of groundwater and their predominant ions are sulfate and hydrogen.

Algae, Bryophyta, Lycopodinales, Equisetales, Filicales, Coniferales, and Angiosperms are all active in the communities living in peat-forming habitats. However, *Sphagnum* is perhaps the most important plant in the development of ombrophilous peats, frequently accompanied by members of the Ericaceae and Monocotyledons, such as *Eriophorum, Schoenus,* and *Trichophorum.* Conifers are also common on ombrophilous peats but they indicate that the water supply to the mire has become deficient which may presage the cessation of bog growth.

Nonsaline rheophilous peats are usually dominated by monocotyledons, including *Carex, Mariscus, Cladium, Juncus, Phragmites,* and *Phalaris.* Bryophyta and algae may be common at the surface of such peats but they rarely contribute substantial quantities of materials, and peats may also be colonized by shrubs and trees, such as *Salix, Rhamnus, Frangula, Betula,* and *Alnus.*

The peat sediment that accumulates in some lakes and ponds, particularly those in swampy areas, is usually allochthonous; partially rotted debris is carried in by streams, and pollen and spores, carried in waters or by the wind, are usually abundant. Algae sometimes make a large autochthonous contribution. That lake peats have been

widespread in the past is shown by the petrographic and paleobotanical
study of coals. Carboniferous seams, particularly in Europe, frequently contain fairly thick so-called "durain" bands. Such wood-
derived material as they contain (20-50%) is highly macerated and
disrupted in various ways. They contain abundant spore cuticles (up
to 70%) and sometimes leaf cuticle and algal remains. It is believed
that for a while the peat became flooded and growth of higher plants
ceased until, through eutrophication, the water became shallow again.
The cannel coals are of this character. The boghead coals were formed
in lakes mostly from algal remains. The extensive Scottish deposit
of torbanite was derived largely from the alga *Bottryococcus braunii*.
The classification and nature of modern peaty deposits in lakes have
been reviewed by Swain [16]. The water in such deposits is usually
nearly neutral.

These theories and characterizations of peat formation and
classification are strongly influenced by European peat formations.
Even within this group of peats, however, the characteristics acribed
to the various classes vary somewhat according to the geologic setting,
as will be seen in a later section.

B. Tropical and Subtropical Peats

Peats in tropical regions have not been studied sufficiently for
one to be able to say whether the above classifications can usefully
be applied to them. Presumably all the peat accumulations in the
Everglades of Florida would be classified on Bellamy's system as
rheophilous. At the same time, it is fair to say that one of the
interests of the area is the intermingling of a number of types of
plant communities with somewhat differing hydrologies [17-20].

Large areas of open water occur in the Everglades, populated by
Nymphaea (water lily) and other aquatics. In the wet season (summer
and fall), there is a slow movement of water from Lake Okeechobee in
the north to the tidal rivers that flow into the Gulf of Mexico.
More or less ovoid islands of *Mariscus jamaicensis* (sawgrass) occur
frequently, oriented in the direction of water flow. The peat surface

in these islands is typically 50 cm higher and the plants offer much resistance to flow, so that the supply of minerals will be less. Islands of cypresses or of mixed hardwood trees *(Persea, Salix, Myrica)* are also common, and these are sometimes found within sawgrass islands. In the latter case, the peat surface is higher still, and because of the presence of the trees as well as an understorey of herbs and shrubs, resistance to flow is also high. The rates of evapotranspiration of the three plant communities are of course different, and this fact contributes to the differences in hydrology [21]. The community that has inhabited a given locality has changed repeatedly during the 2000-4000 years since the peat started accumulating [1] and the changes are reversible. There is little input of ions from the erosion of elevated rocks, but the peats are unusual in that the rock strata upon which they have formed consist of a series of limestone formations. These supply calcium, magnesium, and bicarbonate ions, which buffer the peats close to neutrality, and they are also a source of iron.

Peats that accumulate in saline conditions do not readily fit existing classifications. The nature of the plant associations is determined by the need for salt-tolerance as well as by hydrology. Such peats are common in flat-lying coastal areas in parts of South America and Africa, the lands bordering the Indian Ocean, Malaya, Indonesia, and Borneo, as well as southern Florida. In the estuarine areas and near the coast, mangroves predominate ("mangrove" is the name for a group of about 25 tropical halophilic woody Angiosperms from a number of genera and families).

The coastal area of south Florida has abundant small tidal rivers, inland bays and ponds. Away from the immediate vicinity of these, various monocotyledons are common including *Spartina, Juncus,* and *Mariscus* (in decreasing order of salt tolerance). In all these areas there is a plentiful supply from the saline waters of all needed ions except iron (which, as noted above, can be derived from the underlying limestone). Smith [21] has pointed out the exceptionally high productivity in estuarine environments, resulting from the plentiful supply of ions.

C. Degree of Decomposition of Peats

Many workers, particularly those concerned with putting peat or
peatlands to practical use, also characterize peats by their degree
of decomposition. The various empirical ways of doing this are ex-
cellently reviewed by Farnham [22]. However, because the enormous
Russian literature on peat almost invariably quotes a percentage degree of decomposition, and the basis of the system is hard to track
down, it will be summarized here (from Farnham). This resulted originally from Varlygin [23] in 1924. A smear of peat is placed on
each of three slides, and in ten fields of vision on each slide the
approximate percentage area occupied by structureless brown or black
humus material is estimated.

Most peat beds show a number of distinct layers, and the degree
of decomposition is a useful way of distinguishing them [22]. Peat
accumulation may take place discontinuously, being faster in warm,
dry periods [24], with the interlayers having a higher degree of decomposition than the rest. Rakovskii et al. [25] relate the degree
of decomposition to the concentrations of various peat constituents.

III. PHYSICOCHEMICAL CHARACTERISTICS OF PEAT-FORMING ENVIRONMENTS

A. Introductory Remarks

In mineral soils, ion exchange processes with the surfaces of
clay minerals provide the chief mechanism by which ions are trapped,
held and made available to the micro- and macroflora, though organic
matter makes some contribution. The thermodynamics and surface chemistry of these processes in mineral soils have been extensively
studied [26]. Inasmuch as peats serve as habitats for the higher
plants and microorganisms, there must be here also a mechanism by
which ions are concentrated and made available, and some work has
been done on the relevant physical chemistry, with reference to peats
in fresh water temperate zones. The processes are different in certain respects from those in soils, because of the much higher organic

matter:clay ratio. Humic acids and the amino acids (or peptides) complexed with insoluble organic matter take part in ion exchange processes, and the surfaces of solids in peats must provide extensive areas for adsorption. But an additional factor enters when organic matter in peats has a variety of functional groups so disposed as to be able to chelate metal ions, and some of these chelate complexes are extremely stable [27]. If a metal is complexed in this way it may be unavailable to organisms.

It is the purpose of this section to review what is known about such characteristic properties of peat swamps as water content, pH, Eh, and contents of soluble ions, and of micro- and macronutrients. McLaren [28] has pointed out that the concentrations of nitrogen species in soils are vector quantities; that is, they have not only magnitude but also direction. In appropriate environments, the same is no doubt true of the concentrations of oxygen and of sulfate and other ions. Therefore, data on activities of ions and oxidation-reduction potentials in natural systems containing organisms must be interpreted with caution.

B. Temperature

In some of the northern tundra and muskeg deposits, the peat will be frozen for much of the winter and its temperature may not exceed $15^{\circ}C$ in the summer, except in the immediate surface layers. In the subtropical regions of south Florida, summer daytime air temperatures are commonly $27^{\circ}-33^{\circ}C$ and the peat temperatures $28^{\circ}-35^{\circ}C$, with a decrease of $1-3^{\circ}C/m$ with increasing depth in the peat [29]. Air temperatures in winter may drop to $18^{\circ}-20^{\circ}C$, cooling the upper surfaces of the peat so that at certain times of the year there may be a temperature inversion resulting in some convective circulation of soluble species and also perhaps of free-living microorganisms.

C. pH

The microbial breakdown of lignin is an oxidative process, and there is a tendency for the oxidation of any organic material to produce carboxyl groups. From the generation of humic acids and by other processes, organic acid groups accumulate in the chemical structures of the sediment. However, the actual pH of a peat depends on several factors, one of them being the humic and fulvic acid contents, which vary widely [20] (see Section VI, B).

Under certain circumstances the organic acid groups can be neutralized by metal ions, so that the pH is relatively high. Where the peat is under marine influence, the cations of seawater can exchange with the protons of the acid groups [29]. This will not raise the pH, unless enough bicarbonate is present as buffer. The peats of south Florida are underlain by limestone bedrocks, resulting in a nearly neutral pH. *Sphagnum* peats, which have similar humic acid contents, usually have a low pH (about 3); because of their ombrophilous character they have a poor supply of metal cations.

A limestone bedrock under peat deposits appears to be an uncommon phenomenon, except perhaps in coastal areas; clays, sands, and glacial till are more common; the latter of course can contain a complex mixture of minerals. However, if appropriate rocks near a peat deposit are being eroded, the hydrology of the area may permit an input of calcium and magnesium ions, which together with bicarbonate, may partially neutralize the acid groups and raise the pH. Such conditions are most likely to arise with rheophilous valley swamps characterized by reeds, grasses, sedges, etc. Where the peat is contained in a basin surrounded by higher land the ionic concentrations are higher at the periphery [31-33].

Lowmoor peats tend to have the highest inorganic content and highest ph [30, 34]. Thus, Szmytowna and Latour [35] studied ten peat deposits and found the inorganic content to be 2.5-4% and the pH 4.2-5.1 from "high" beds, and 9-23% and 5.6-6.4, respectively, in "low" beds. However, Volarovich et al. [13] give wider ranges for pH from a study of 288 samples from many deposits: 2.8-4.4 for highland peat, 3.2-5.3 for intermediate peat, and 2.8-7.4 for lowland peat. The wide range for the latter category is noteworthy and illustrates the point that the input of ions depends on the geologic setting. Acidic lowland peats are relatively uncommon, however, and

a supply of ions sufficient for more or less complete neutralization
is usually available.

Peats have a well-developed fibrous and/or colloidal structure
that holds water with varying degrees of affinity. Thus, peats to
some extent create their own water table; by surface tension they can
draw up water to remain saturated, though no doubt high rainfall is
needed to stabilize the situation. W. G. Smith (personal communication, 1968) has noted mangrove swamps in Malaya in which the peat
surface is permanently some way above the level of water in adjacent
streams, and yet the trees must be growing in at least moderately
saline conditions. Hence, there will be some supply of nutrient ions
even to raised or ombrophilous bogs, by diffusion from ground waters
at the base of the profile.

D. Oxidation-Reduction Potentials

Few measurements of oxidation-reduction potentials [Eh] have
been made in peats. Szilagyi [36] has measured Eh in buffered humic
acid/water suspensions and believes that the electrochemical ion
equilibrium in the system humic acid-water is maintained by the par-
ticipation of one proton and three electrons. Measurements of Eh
have been made on freshly opened cores from two Minnesota peat bogs
[37, 38], but it is not clear to what extent they were affected by
oxygen absorption. In the Rossburg bog a steady drop with increasing
depth was found, from +400 mV near the surface to +200 mV at 4.6 m
(over the same range the pH rose from 3.19-6.5), which in the Cedar
Creek bog the values were in the range +300 to 250 mV (and pH 7.0-
7.5).

Armstrong [39] studied a blanket bog peat in Sutherland (Scotland) and a valley bog in North Yorkshire (England), using as a probe
for *in situ* measurements a platinum wire electrode mounted in the end
of an 80-cm rod. Only plots of Eh against oxygen diffusion rates,
measured coulometrically with the same electrode, are shown. Values
in the blanket bogs ranged from about +590 to +70 mV and from +690
to +60 mV in the valley bog (all values corrected to pH 6 on the

assumption of ΔE = 60 mV per pH unit). Measurements with a similar probe were made in peat cultivated for wheat, near Lake Balaton in Hungary [40]. Except in the period immediately after the area had been flooded, the Eh dropped fairly steadily from +680 mV near the surface to about 0 mV at 95 cm, with a small reversal of the trend between 35 and 45 cm where there was a layer rich in ferric iron. In spite of the positive potentials, the layers from 55 cm down were believed to be permanently anaerobic.

Eh profiles at ten sites in the peat bogs of southern Florida, measured with a pyrolytic graphite electrode mounted in a probe, were all in the so-called oxidizing range (+500 to +25 mV). In the fresh-water environments the value was almost independent of depth, but in saline environments there was a considerable decrease in Eh with depth. In the saline peats the change with depth may be associated with bacterial reduction of sulfate ion. However, the Eh values were much more positive than would be expected from a system in which equilibrium between H_2S and SO_4^{2-} determined the measured potential [29].

Therefore, the measurements available from several different environments agree that the Eh of peats is positive even at considerable depth. Such results have sometimes been interpreted as indicating aerobiosis in the microflora but this view cannot be accepted. It is true that roots may penetrate to some depth and permit some aerobic activity in the rhizosphere, but such microenvironments cannot be studied by normal electrode methods. Anaerobiosis is indicated in saline peats by the evidence that sulfate reduction takes place and perhaps also by the very sharp drop in aerobic microorganism populations with increasing depth (see Section V, D). Organic electro chemical processes may contribute to the measured electrode potential but be irreversible. The activities of many inorganic species (O_2, forms of N, SO_4^{2-}) at any given depth will depend on the rate of growth of the elements of the microflora. Consequently, species that affect the Eh electrode may have steady-state concentrations but can hardly be in equilibrium in the systems under discussion. Therefore, as Whitfield [41] has suggested, Eh measurements may be regarded as

useful operational parameters characterizing an environment but cannot be considered as redox potentials of thermodynamic significance
(see also Stumm [42]).

E. Water Contents

The structure of peat is extremely complex, and this must have
a bearing on the biochemical processes occurring. Any plant tissue
is highly swollen, and intact cells will be filled with immobilized
water. The spaces between fragments of tissue and/or particles will
often be narrow and capillary forces active. In addition, there is
fine-grained material of colloidal character for which the degree of
dispersion will depend on pH, ionic strength, and the nature of the
ions. There are, of course, various ion-exchange and adsorption
phenomena. Excellent reviews of these problems have been presented
recently [43, 44].

Various methods have been used to distinguish between the different ways water is held, most of them depending on radioactive
tracers (^{45}Ca, ^{35}S, ^{131}I, ^{14}C, and ^{3}H). The adsorption of ^{45}CaCl
and $Na_2^{35}SO_4$ has been used to distinguish between free intercellular
water and structural gel-bound (or immobilized) water, the latter
having no dissolving capacity for ions [45]. The $3H_2O/^{35}SO_4^{2-}$ ratios
in the intergranular water have been compared [46] in various model
systems: sand, thin glass needles, silica gel, and mixtures of peat,
sand, and silica gel. Comparisons of the ratios showed that in the
dispersions of porous materials, a considerable part of the water
was inaccessible to sulfate ion.

Peats contain slit-shaped pores or capillaries 0.3-20 μm in
width [43, 44]. The kinetic surface area is 1.5-28 m^2/g for high moor
peats and 2.7-7 m^2/gm for low moor. Owing to the high content of
static water the effective porosity is only 0.4-0.6 instead of the
0.85-0.92 one might expect from the total water content. There are
diffuse regions around the separate electrically charged surface
groups, but there is no continuous electrical double layer. This is
said to explain the very low electrokinetic potential (<3-15 mV).

Measurement of desorption isotherms indicate a total evaporation energy of more than 15 kcal/mole (cf. 9.7 kcal/mole for pure water).
Changes in NMR linewidth, and in the spin-lattice and spin-spin re-
laxation times generally confirm the above conclusions on partially
immobilized water [44].

Korol [47], in a comprehensive study of various peat types occurring in northwest Russia gives the "chemically and physically
bound" water as in the range 38-56%, and "strongly bound" water as
9-13%. Voralovich and Churaev [43] quote values for "immobilized
water" of 67-82%.[*] They also found that an increase in concentration
of Ca^{2+} or other polyvalent ions tends to convert the peat colloids
to a coagulated gel structure that immobilizes much water, whereas
an increase of Na^+ or K^+ tends to give sols or true solutions, thus
liberating water. In later work [48] it was found that elution of
ions with deionized water caused structural rearrangements and a
tenfold decrease in flow rate, whereas elution with NaCl solutions
(0.001-one mole/liter) increased the flow rate.

These data imply that the diffusion of soluble species, including
nutrient ions, will be considerably slower in peats than might be expected from their high water contents (usually 80-95%). The transport
of ions in peats has been described as a gel diffusion process [49].
It is also implied by the above discussion that the physical structure
of saline peats will have some characteristics not found in freshwater peats.

F. Ion-Exchange Phenomena

The total ion-exchange capacity of peats is in the range 20-200
mEq/100 g dry peat [43, 50, 51] (the measurements are made without
drying, since the capacity is thereby reduced by 20-70%). The exchange
is reversible [50]. Bel'kevich et al. [50, 52] measured the ion-exchange equilibria of a variety of mono- and divalent ions with the
acid form of peat at various temperatures ($25°$-$75°$C). The free-energy

---

[*]All water contents are given on the wet basis.

change in the exchange of monovalent ions was +1320-2350 cal, whereas for divalent ions it was only +190-280 cal. Since they found that
the exchange process occurred with no change of enthalpy, it appears
that the exchange of of H^+ for metal ions increases the entropy of
the system. The extent of exchange at equilibrium increased in the
order $Li^+ < Na^+ < K^+ < Rb^+ < Cs^+$, and $Mg^{2+} < Ca^{2+} < Cd^{2+} < Ba^{2+}$.
Salmon [53] found that Mg^{2+} was held by an English fen peat much less
strongly than Ca^{2+}.

The extent of desorption of certain ions (Ca^{2+}, Mg^{2+}, Fe^{2+}),
by one N KCl was found to be 86-90% of that produced by 0.1 N HCl
indicating that the ions were partly in the form of relatively stable
organic complexes [54]. The formation of chelate complexes of Fe^{2+},
Cu^{2+}, Zn^{2+}, and Mn^{2+} was deduced from potentiometric, polarographic,
and spectrophotometric studies of peat humic acids [55]. Similar
conclusions relating to these and other metals have been discussed
by other authors [27, 30]. Electrodialysis of the centrifuged waters
from bogs showed that the organic particles and part of the metal ions
were negatively charged, migrating to the anode, presumably as humate
complexes. In particular, much of the iron was held in this way [56].

In a recent study of remarkable comprehensiveness, the ion exchange behavior of carefully purified fulvic, hymatomelanic and humic
acids from a Hungarian lowland peat was studied [57]. The exchange
capacity for H^+ was greater for the fulvic and hymatomelanic acids
(800-900 mEq/100 g) than the humic (400-600), and was also greater
than the capacity for a series of 12 polyvalent cations, mostly of
transition metals. The range of capacities found was wide, ranging
down to 50-150 mEq/100 g for Mg^{2+} and zero for Cr^{3+}. Estimates of
the partial ionic character of the metal-acid bond were obtained
from the frequency of the $-COO^-$ vibration (1585-1660 cm $^{-1}$), epr
measurements and Mössbauer spectroscopy. Thus, the bonds formed by
Fe^{2+} were found to be 26% and 37% ionic in humic acid and fulvic acid,
respectively. Bel'kevich et al. [49, 58] found the rate of exchange
of various ions with peat to decrease in the same order as the equilibrium extent of exchange increases (see above); thus, for example,

little Li^+ exchanges, but equilibrium is reached rapidly. The activation energy for Ca^{2+} exchange was found to be 9.11 kcal/mole. It
was these results, and the dependence of rate on particle size, that
led the authors to conclude (as noted above) that ion exchange in
peats is a gel diffusion process.

It was pointed out earlier that the actual amounts of ions in
peats as they exist naturally depends on the type of swamp considered.
Large volumes of data on the concentrations of the common cations in
many peats have been accumulated by some authors [32, 44, 47, 59].
Mornsjo [60] found characteristic differences in the distributions
of Ca, Mn, Fe, and Al between rheophilous and ombrophilous peats,
and in the Ca:Mg ratio, although some of the latter type did not fit
the common pattern, perhaps because of secondarily supplied minerals
brought in by man.

The maximum concentrations of Ca and Mg (in the rim and bottom
layer of peat beds) were thought to be controlled by incoming waters
and the geological environment. On the other hand, the concentration
of Mn was affected by biological factors; it was highest in the upper
peat-forming layers [32]. A biological influence on total mineral
matter contents has been noted [61], which were highest under sedge-
grain or reed-sedge phytocenoses with certain mosses and willow bushes,
and lower under sedge-grass phytocenoses with horsetail and elder
bushes.

There is a marked tendency for higher concentrations of cations
to be found in peats of higher pH. According to Volarovich et al. [13]
the mean correlation coefficient for Ca^{2+} in 288 peat samples, for
example, was 0.84, and 0.91 for $Ca^{2+} + Mg^{2+}$. The characteristic pH
ranges of rheophilous, transitional, and ombrophilous peats are different, as has been noted, and in accord with this the average contents of $Ca^{2+} + Mg^{2+}$ in peats of these types differ (see above).

The mode of combination of iron in peat has been studied more
than that of other metals. Thus an ombrophilous peat had 2.15% Fe
as silicate, 31.3% in humic complexes, and only 0.05% in ion-exchange-
able form, and these data were believed to be typical [62, 63]. In

contrast, rheophilous marshes have most of their iron in inorganic form (hydroxides and in some cases phosphates and carbonates). Differences in iron-humic complexes have been associated with differences in the contents of various oxygen functional groups [64].

G. The Fixation of Sulfur in Insoluble Forms

The occurrence of iron in peats is to some extent associated with the fixation of sulfur. The Carboniferous coals of the Interior Province of North America (e.g., those of Illinois, Indiana, Kansas, and Missouri) are high in sulfur, as also are some Appalachian coals. However, the Cretaceous and Tertiary coals west of the Mississippi are consistently low in sulfur. Paleontological studies of the associated strata show that the western coals were all deposited in fresh water conditions, whereas those of the Interior Province and parts of Appalachia were laid down under conditions of marine influence. That is, there is a correlation between high sulfur contents of coals and the incursion of saline waters into peat swamps. This correlation is found today in the peats of southern Florida [20, 21, 65], where saline peats have sulfur contents of 2-6% (dry basis) and freshwater peats typically 0.1-0.6%.

This empirical correlation strongly suggests that the much higher concentration of sulfate ion in marine than in fresh waters is a key factor, and that the starting point for the conversion of sulfur to insoluble forms is bacterial reduction by *Desulfovibrio* spp. or *Clostridium* spp. The two principal forms in which sulfur accumulates in peats are as pyritic and organic sulfur. Iron is precipitated by H_2S as troilite, FeS, and reaction with elementary sulfur is required to convert this to pyrite, FeS_2. Given and Miller [65] have found elementary sulfur (0.01-0.2%) in saline peats, and have discussed how reduction of sulfate to H_2S and oxidation of H_2S to S may occur in the same gross system. They naturally invoke separated microenvironments, and envisage two possible oxidation mechanisms: in the rhizosphere zones, where oxygen will be available from the roots, and anaerobically by *Thiobacillus denitrificans* in the

presence of nitrate ion [66]. Cooper [4] has found a small amount
of elementary sulfur in the secondary roots, presumably in the cor-
tical tissue, of a specimen of *Rhizophora mangle* L, the red mangrove,
growing in the Florida Everglades. There is no evidence to show
whether or not *T. denitrificans* is active.

H. Trace Element Distributions

There have been a number of studies of the distribution of trace
elements in peats [20, 31, 59, 67-72] and some [70, 72, 74] distinguished the available amount as a fraction of the total present. The
contents of such elements as Zn, Mn, Mo, Cu, Cr, and Ni are highly
variable in different peat deposits and seem to be determined by the
nature of the rocks around the peat basis, the underlying bedrock,
and the groundwater hydrology [31, 33, 70, 72].

In three marshes on the Baltic Shield in northwestern Russia,
which were surrounded by highland containing ultrabasic rocks (low
in SiO_2) high contents of Cu, Ni, Co, and V were found, whereas in
other marshes surrounded by arenaceous-argillaceous (sandstone-clay)
rocks, the contents of Be, Ga and rare earths were particularly high
[33]. Similar differences in Cu, Zn, and Mn were found for other
basins according to the nature of the bedrock. It was also noted
that the total contents of these ions and the mobile (or soluble
fraction of them) varied widely with the seasons; in May, the mobile
fraction of all three ions was 40-80% of the respective totals, whereas at the period of minimum mobility and total content only 10-25%
was mobile. These observations presumably reflect large inputs of
ions in solution at the time of spring influx of waters from higher
ground.

The ability of a Welsh sedge peat to retain strongly held Cu when
acidified was reduced to about half if either the carboxyl or hydroxyl
groups, or both, were converted to nonpolar derivatives, pointing to
chelation by adjacent COOH and OH (or quinone and OH) groups as the
retention mechanism. The retention of Cu by whole peat was greater
than retention by the extracted humic acids [71].

There do not seem to be good correlations between trace element concentrations and the nature of the higher plant populations that gave rise to the peat [67]. However, the presence of some metals at finite concentration is no doubt to be attributed to the needs of the macrophytic population, while the presence of Be is unconnected [73].

The components of peat undoubtedly have the ability to hold strongly and concentrate a large number of metals from ground waters [27]. The fact is that, generally speaking, they do not. The concentrations reported by all the authors referred to [31, 33, 59, 67, 70, 72, 74] are below, usually well below, the average abundances in the earth's crust, with the exceptions that manganese [69] and molybdenum [31, 75] may sometimes exceed their mean geochemical abundances. The situation appears to be that peats can indeed concentrate trace elements by a large factor relative to the incoming ground waters, but the supply is usually not enough for the process to yield concentrations above the Clarke value. The interaction of trace elements with organic matter has been generally reviewed by Manskaya and Drozdova [20], who also give further data on peats.

Is the balance between productivity and decay of biomass in peat swamps in any way controlled by the supply of micronutrients? For Cu, Zn, Mn, and Co, only small proportions of the metals present in peats of the Novgorod region are available to organisms, although essentially all of the Mo is available [74], elsewhere lower proportions of the Mo are available [75]. From the point of view of those who wish to convert peatlands to the production of food crops, peats are notoriously deficient in Cu, and sometimes other metals [27, 76]. However, the plants indigenous to swamps no doubt have more modest requirements. If the productivity of the macrophytes were limited by these factors, the growth of the microflora would have to be even more restricted to permit accumulation of organic debris. However, there does seem to be a possibility that the growth of some elements of the microflora may be limited in peat swamps by the low availability of certain essential micronutrients, but Mo, a constituent of both nitrate reductase and nitrogenase, does not usually seem to be in short supply.

I. Availability of Macronutrients

Few studies of the nutritional aspects of peat accumulation seem
to have been made, if one wishes to consider the nitrogen and phos-
phorous economy of the indigenous flora and microflora instead of the
use of the peat for agricultural purposes.

In two peats of northern Russia (forest steppe and sedge peats)
the total nitrogen contents of the top 20 cm of the drained but not
cultivated material was reported as 2.4% and 3.0%, whereas the nitrate
contents were only 16 and 4 ppm, respectively [77]. These data were
interpreted [78] to mean that at any one time there is no more than
2.3 kg/ha available nitrogen. However, how much of the nitrogen in
peats is available seems to be a matter of some controversy [79].

In various peats in Poland, 97-99% of the total N in low moor
and 89-96% in high moor peats is present in organic form [80]. The
process of peat formation from plants was said to result in a decrease
in amine N and a corresponding increase in nonhydrolyzable amide N.
(As will be discussed in Section VI, D, we do not believe this to be
generally true; the concentration of free amino acids is no doubt very
low in peats, but the total concentration releasable by acid hydrolysis is usually high compared with that in the higher plants). The
formation of nitrate ion was observed, particularly in surface layers.

In contrast, no nitrate ion and no evidence of the nitrification
of ammonia to nitrate ion was detected in two Irish peat bogs (a blanket and a raised bog), even after drainage and fertilization, unless
the peat was also limed and left for several years [78]. No denitrification was observed unless nitrate was added to the system together
with some lime to raise the pH. Similar observations were made on
two Polish peats [81].

It seems to be generally true (Section V) that peats contain very
limited populations of both aerobic nitrogen-fixing bacteria and nitrifying bacteria, though some anaerobic nitrogen fixers are found.
Thus, empirically we can say that productivity and decay rates in
peat swamps might be somewhat limited by nitrogen supplies.

The situation is somewhat different in subtropical and tropical

environments. Blooms of members of the Cyanophyta are common in
freshwater swamps in South Florida, and they can materially increase
nitrogen availability; yet the peats still accumulate, which suggests
that the growth of microbial destroyers of plant tissue is not limited
by nitrogen deficiency. The productivity of mangroves and other
halophytes in estuarine environments is well known to be exceptionally
high [21]; seawater does contain some nitrate, but the average content in ocean water is only 0.001-0.7 ppm. Some preliminary measurements of the nitrate concentrations in saline and freshwater Florida
peats, using selective ion electrodes, gave values of about 3-30 ppm
(P.H. Given, D.J. Casagrande, and A.J. Lucas, unpublished observations,
1968, 1972).

IV. PRESERVATION OF CELLULAR TISSUE IN PEATS

The extent of the preservation of plant tissue is made use of
in several methods for estimating the degree of decomposition of
peats (see Section II, C). The use of "botanical indices," to be
used with chemical characteristics, has been proposed for the
classification of bogs [82]. These uses depend on a quite unsophis-
ticated view of the preservation of macrophyte tissue. We are concerned here only with such information from more informed studies as
may bear on the biochemistry and microbiology of peat.

The use of pollen analyses in profiles of sediments of various
kinds to permit reconstruction of past climates and synecologies is
well known, and this approach has been used to some extent with peat
swamps [83-86].

Many of the studies of plant fragments in peats have been carried
out in research designed to elucidate the origin of coal macerals or
phyterals. Under the microscope, all coals appear heterogeneous to
a greater or lesser extent, and some of the entities seen are clearly
identifiable with plant organs or parts of organs, e.g., xylem tissue,
the cuticles of spores, pollen, and leaves, and sometimes parts of
the leaf tissue itself. Other entities are amorphous, and are assumed

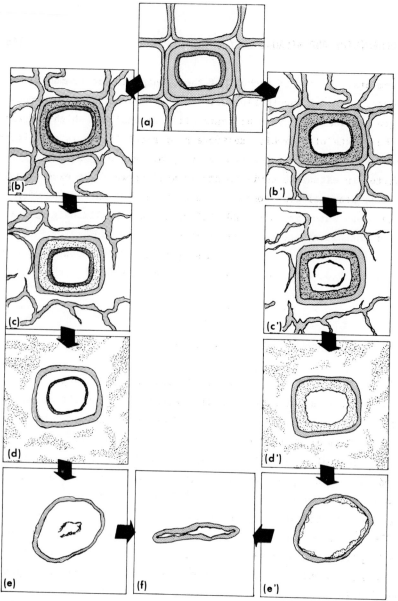

FIG. 1 Two alternative paths of alteration of the secondary cell walls of the sclereids of *Nymphaea* roots and rhizomes [1]. (a) Schematic diagram of an unaltered *Nymphaea* sclereid with a three-layered wall structure; (b) Schematic diagram illustrating the first step along one of the paths of alteration. The middle layer of the wall becomes granular; (c) The step after that shown in stage (b); only a few grains of the middle layer remain; (d) The step after that shown in (c). The middle layer is gone and the outer layer has become thinner; (e) The step after that shown in (d). Only a few fragments of the outer layer remain; (f) The step after that shown in either (e) or (e'): all layers but the inner layer (the primary wall complex) have disappeared and the cell has collapsed; (b')

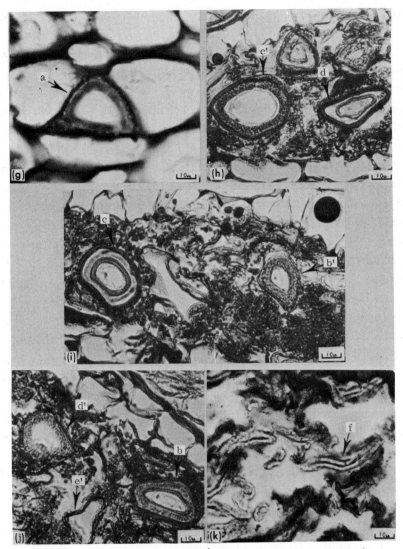

cont.

Schematic diagram illustrating the first step along a path of alteration which differs from that started at (b). The outer wall layer becomes thin at the same time that the middle layer becomes granular; (c') The step after that shown in (b'). Only a few fragments of the outer layer remain, but the middle layer has not changed greatly from that of stage (b'); (d') The step after that shown in (c'). The outer layer is gone and the middle layer is noticeably thinner; (e') The step after that shown in stage (d'). Only a few remnants of the middle layer remain; (g) Photomicrograph comparable to (a); (h) Photomicrograph illustrating stages of alteration comparable to stages (c') and (d); (i) Photomicrograph illustrating stages of alteration comparable to (c) and (b'); (j) Photomicrograph illustrating stages of alteration comparable to (d'), (e'), and (b); (k) Photomicrograph illustrating an alteration stage comparable to (f).

to represent plant tissue so altered by biochemical processes that
all morphology has been lost. In immature coals, fossil tissue may
be so well preserved that the genera of plants and the nature of the
tissue can be easily identified, and differential alteration of the
cell walls of some tissues can be contrasted with almost intact preservation of others [87].

It has been known for some time that some tissues are more resistant to decay than others, and that even within the same fragments
of apparently homogeneous tissue differential decomposition may occur
[88-92]. As a result of his study of woody fragments in Massachusetts
peats, Barghoorn concluded that the various layers of a mature scler-
enchymatous cell wall decomposed in the sequence illustrated in the
series (a)-(f) in Fig. 1. In this sequence the inner layers of the
wall become detached from the outer during the decay process. Cohen
[1], in his study of Florida peats, observed this mode of decay in
the walls of sclereid cells of the roots and rhizomes of *Nymphaea*,
the common water lily, but he also observed sometimes another mode
in tissue of the same type, also illustrated in Fig. 1 in which the
innermost layer was attacked first and decay proceeded outwards.
Cells representing all the stages in the two alternative paths of
degradation could be found within a few microns of each other; see
Fig. 1, (g)-(k).

There is of course a different distribution of lignin, cellulose
and the hemicellulose polymers in the various layers of the wood cell
walls, and there are differences in crystallinity and the orientation
of the crystallites. There must be biochemical explanations for the
order in which the layers are attacked and for the existence of the
two modes, although it is not obvious what these might be.

Other authors have considered plant tissue in peats from the
point of view of coal petrography [2, 93, 94]. The article by Jacob
and Koch [2] is of interest in pointing out the different microscopic
techniques appropriate to the examination of different kinds of plant
parts. Thus, reflected light fluorescence microscopy distinguishes
particularly well among pollen and spore exines, cuticle, suberinized

cell membranes of cork tissue, and "polymeric-bituminous algal material." Fluorescence techniques are also useful in examining woody tissues, particularly those with inclusions of humic gel and of phlobaphenes. In the examination of woody cell walls, transmitted light techniques are preferred, using polarized and unpolarized light. Particles of charcoal, from natural fires, and fungal remains can best be detected by reflected light, because they have higher reflectances than other organic components. This type of study also lends itself to estimates of the degree of decomposition [2, 83].

Given [95] has observed, using polarized light microscopy, that severely disrupted tissue may still retain cellulose in what remains of the cell walls, since the crystallinity is still evident from their birefringence. In contrast, the cell walls of fairly intact tissue may show very little birefringence, indicating that cellulose can be removed without destruction of the cell morphology. Spackman and Barghoorn [87] found that the cell walls of the secondary xylem of *Persea* retained cellulose, as evidence by birefringence, even after coalification, in their study of the lignite from Brandon, Vermont.

A novel approach to the problem of the alteration of plant remains during peat formation was the use of electron microscopy to examine the changes in the submicroscopic features of leaves [96]. The results of this study suggested that most major alterations in the cuticularized leaf surfaces take place at the litter stage before the material becomes part of the peat proper. Other authors [97-99] have also used the electron microscope to explore aspects of peat degradation and microbiology. A further advantage claimed is that the morphology of microorganisms can be more clearly seen than with optical methods.

Grosse-Brauckmann and Puffe [83] used both pollen analysis and microscopic examination of thin sections in an interesting study of profiles of a peat swamp in northwest Germany. They were able to reconstruct the vegetational history of the area, showing a transition from a reed-sedge population through a forest swamp with birch trees to a high moor type with *Sphagnum* and pine trees. Fruits, seeds, stumps of wood, and heather organs are well preserved, but the parenchyma of the plant organs had largely disappeared. Retention of cellulose by well-preserved woody tissue of the leaves and stems of heather, herbs and arborescent plants was demonstrated by microreactions and by optical methods. However, strongly humified tissue is often free of cellulose. Fungal hyphae and fecal pellets of the peat microfauna were often seen.

The most comprehensive investigation of this kind appears to be that by Cohen [1], who studied the peat-forming environments of the Florida Everglades, by comparing oriented thin sections of peat with sections of the fresh organs of relevant plants. Vegetational histories in the different areas could be reconstructed even when only greatly altered fragments of organs were present.

The transformation of particular tissues and their cells as they become peatified were described. Cohen found that a relatively small percentage of any of the peats studied was stem wood as compared with that part representing roots or leaves; the importance of root decay in peats has not been stressed before. In some layers of mangrove peat, however, no leaves were found, since tidal action in some areas removes leaves and other small litter from the surface. He also reports the percentage of cells and small cell fragments (<100 μm), and notes that it is high for freshwater peats and variable for the saline mangrove peats. Thus if stem material makes a larger contribution to peat than is indicated by the easily recognizable pieces, its contribution must be to the cells and small cell fragments and/or to the amorphous humic matter. Unlike Grosse-Brauckmann and Puffe [83], Cohen did observe parenchymatous cellular tissue of various kinds in peats, and indeed sometimes found this type of tissue (with thin cellulosic cell walls) to survive in plant parts in which tissue with lignified cell walls had already disappeared. Four genera and species of mangroves are found in the coastal swamps of Florida, and three of these are notable for having large numbers of colored cell fillings in parenchyma and other cells of most organs of the plant. The inclusions take the same stain as lignin. Mangroves are well known as rich sources of tannins, and indeed are used as commercial sources of materials for tanning hides. It appears very probable that the colored cell fillings are partially polymerized leucoanthocyanins (i.e., condensed tannins).

One mangrove, *Avicennia germinans*, has been observed to contain few of these cell fillings and a very small amount of chemically extractable tannins. Cohen frequently encountered a peat type to which both *A. germinans* and *Rhizophora mangle* had contributed root material, but leaf and stem material from *R. mangle* only was found. He suggested that the tannins of *R. mangle* exercised some protective action on its aerial parts during decay in surface litter and incorporation into the sediment, whereas aerial parts of *A. germinans* were not able to enjoy such protection.

By identifying the genera and species that have contributed to a peat, the botanist is drawing attention to the plants the chemical components of which should be compared with what is found in the sediment.

V. PEAT MICROBIOLOGY

A. Introduction

The question as to why plant debris is incompletely destroyed by microorganisms is essentially relevant to the surface horizons of peat, to the accumulated litter that is not yet peat, and to the senescing plant tissues above or, in the case of roots, within the peat. Although it is difficult to determine accurately the actual horizon in which the litter becomes peat, it is nevertheless important to consider the microbial decomposition of the litter of peat-forming plants as compared with situations where the destruction of organic matter is approximately equal to its production. Microbial growth within the peat mass may alter the chemical or physical nature of the sediment which may be of significance in considering its long-term future. Activity in similar sediments during the coal-forming periods might have resulted in profound differences in the chemistry of various strata in a seam, which in fact are not found at least in those parts of seams primarily derived from woody tissue. Changes in peat chemistry down the profile may affect its properties if it is reclaimed for agriculture or used in industrial processes. Such microbial activity may also selectively alter the historical record of pollen and macroscopic plant remains [100, 101]. It is a basic tenet of such research that the remains have not selectively decayed and if microbiological activity is shown to occur within peat, then data from pollen studies must be viewed with great caution.

Reclamation or major alterations leading to degeneration of peatlands appear to be a necessary part of most agricultural economies based on this soil type. Reclamation varies from the drainage or afforestation of virgin bog to the use of cut-away peatland following the removal of the bulk of the peat for fuel. In all these processes, and considering the use of peat as a soil additive, it is pertinent to ask whether, in the changed environmental conditions, microbial decomposition will recommence to such an extent that the peat will thus be ultimately destroyed? Such microbial activity would be significant when considering the question of the depth of peat to be left following fuel cutting, and it affects the economics of mineral soil improvement processes that involve peat additions.

B. Microbial Aspects of Peat Formation

Amongst the several possible approaches to the question of peat formation, perhaps the most obvious method is the study of the litter of peat forming plants, to follow its decay as it is incorporated into peat [102]. However, in many studies the surface features of the peats are poorly described and it is difficult to determine whether the "surface" samples refer to the formative peat horizon or to its subsequent altered product.

Latter and Cragg [103] found that *Juncus squarrosus* leaves on the surface of an ombrophilous peat deposit lost 20-30% of dry weight per year in the first 2 years following senescence. During this

period a number of microfungi were common on the leaves but as they
became buried deeper in the H_1 horizon, the amount of fungal mycelium
on the litter decreased sharply. Relatively few bacterial species
were active on the decaying leaves, but several types of animal were
recorded on or in the tissue. Clymo [104] has shown that most of
the decomposition of *Sphagnum* in ombrophilous bogs takes place in the
surface horizons. Samples buried in the surface and at 75-cm depth
were decomposed in the ratio of 13:2 (estimated as dry weight loss),
respectively. Animal grazing was discounted as an important factor
in decomposition and decay was not accelerated by additions of either
inorganic salts or peptone.

The absolute size and qualitative composition of the surface
populations of microorganisms may give an indication as to the nature
and extent of the decomposition processes that affect litter at the
peat surface, and comparison with those in natural or agricultural
loam soils or with those deeper in peats themselves may provide valuable indications as to the nature of the surface activity.

Kox [105] reported that pure cultures of several cellulolytic
fungi and bacteria, which were also in some instances pectinolytic,
were unable to decompose *Sphagnum*. Mixed cultures increased the
extent of moss decomposition, perhaps through the combined effects
of several enzymes. The resistance of *Sphagnum* tissues to decay may
be partly due to the complexity of its cell walls, which contain
cellulose with a high proportion of xylan units, a high pectin content, and a phenolic compound termed "sphagnol." Further chemical studies of *Sphagnum* decomposition indicated a reduction in the cellulose
content of the tissues from 35% in the living moss to 10% deep in
raised bog (ombrophilous) peat, and 8.5% in blanket bog peat [106].
Chastukhin [107] demonstrated that several saprophytic fungi isolated
from the surface layers of an actively forming peat bog were relatively ineffective in decomposing *Sphagnum*. If, however, the moss
was inoculated with *Collybia dryophilia*, a characteristic species of
drier peat bogs, then as much as 60% of dry weight was lost, suggesting
that the microbial populations of ombrophilous peats are unable to
destroy the litter of the peat-forming plants.

C. Microbial Populations in the Surface Horizons of Peat

Numerically the most abundant group of organisms in peat are the
bacteria. The populations recorded have varied widely, however, depending on the type of peat examined and the counting methods employed.
Methods that involve culturing suffer because of the inherent selec-
tivity of every known growth system, and direct observation techniques
have the disadvantage that both live and dead cells may be recorded.

Using relatively nonselective media and aerobic incubation conditions, bacterial counts of 340 x 10^3 to 55 x 10^6 per gram dry weight
(which may not be the most appropriate method of expressing the data
in comparison with mineral soils) have been obtained by a number of
workers for the superficial horizons of ombrophilous bogs [108, 110).
Much higher counts were obtained by Latter et al. [111], who used a
Jones and Mollison direct observation method that gave figures for
Juncus peat of 1.8-3.9 x $10^9/cm^3$. Fewer studies have been carried
out on rheophilous peat. Visser [112] recorded 270 x 10^6 bacteria
in the surface layers of a tropical *Papyrus* peat, and we have obtained
values ranging from 22 to 52 x 10^6 per gram dry weight for freshwater
Everglades peats. In saline mangrove peat bacteria were less numerous, with 6.2 x 10^6 per gram dry weight being recorded at various
points along the seaward margin of the Everglades.

Waksman and Stevens [113] demonstrated that there were considerable numbers of fungi, aerobic cellulolytic bacteria, and nitrifying
bacteria at the surface of low moor (rheophilous) peat as compared
to the lower horizons. In contrast, ombrophilous peats had low numbers of bacteria in the surface layers with an increasing population
further down. They also carried out some experiments to determine
the extent to which *Cladium (=Mariscus)* and *Sphagnum* litter could
be decomposed and they found that the rate of breakdown of fresh
litter was faster than that of the peat formed from similar litter.

Noteworthy qualitative aspects of the bacterial flora include
the general paucity of several important physiological groups. Such
peats as have been studied have very few aerobic nitrogen-fixing
bacteria [114, 115]. This seems surprising because a variety of

microorganisms inhabit root nodules, forming a symbiotic relationship with many nonlegumes, including members of the Myricaceae, Coniferales, Cycadaceae, and of the genera *Alnus* and *Ginkgo*, all of which are peat formers now or have been in the past [115]. Although the counts of nitrifying bacteria are variable [113], they seem to be scarce, especially in ombrophilous habitats [116, 117]. Tesic et al. [118] report the presence of aerobic ammonia formers in peat. Aerobic cellulose decomposing bacteria are uncommon in ombrophilous peats, but Visser [112] recorded 44 x 10^3 per gram dry weight in the *Papyrus* peats. Paarlahti and Vartiovaara [108] noted that sulfate reducers and thiosulfate oxidizers were present in several acid peat types but oxidizers of elementary sulfur were more restricted in their occurrence. Anaerobic species, including some that can fix atmospheric nitrogen, occur in the surface layers of peat although they undoubtedly become relatively more important in the deeper strata.

Amongst the bacterial genera recorded, *Bacillus*, *Pseudomonas*, *Achromabacter*, and *Arthrobacter* have been frequently noted in ombrophilous peats, and *Bacillus* and *Clostridium* are commonly encountered in rheophilous systems. *Cytophaga* species were found to be quite abundant in peat as they are in Brown Earth loam soils. Occasional records of *Streptomyces* have been reported from ombrophilous peats [119]; some *Micromonospora* and *Nocardia* isolates have been found in an extensive study of Japanese peat soils [120], and Zimenko [114] has recorded several species of *Actinomyces* in both highland (? ombrophilous) and lowland bogs. Populations of Streptomycetes in Everglades peats varied from 3 to 286 x 10^3 per gram dry weight in freshwater sites, and from zero to 17 x 10^3 per gram dry weight in saline habitats. Skrypka and Lysenko [121] noted that many of the actinomycetes they isolated from peat soils, mainly species of *Streptomyces*, were antagonistic toward gram-negative bacteria.

Microfungi are usually abundant in the surface layers of peat and they may be significant here because of their primary importance in litter decay. As with the other two groups of organisms, the size of populations recorded varies enormously. In ombrophilous peat

Holding et al. [109] recorded 10-70 x 10^3 propagules per gram dry
weight, and Baker [110] counted 2.9 x 10^6 yeasts per gram dry weight.
Latter et al. [111], using a direct examination technique, gave estimates of 160-320 m/cm^3 of fungal hyphae for two acid peats. As is
usual with such methods these results indicate the presence of far
larger populations than were recorded by culturing techniques.

   In rheophilous peats, the numbers of fungi recorded are frequently
higher than in ombrophilous habitats. Visser [111] found that the
surface of *Papyrus* peat had a fungal population of 5.8 x 10^3, and our
own data show that freshwater Everglades peat possessed 27.4-174 x
10^3 per gram dry weight. Saline Everglades peat had a slightly smaller
population of 0.8-40.2 x 10^3 per gram dry weight.

   More is known concerning the nature of the fungal populations
because of the comparative ease with which these organisms may be
identified. Acid peats are characterized by species of *Penicillium*,
Cladosporium, *Trichoderma*, *Mucor*, *Mortierella*, *Cephalosporium*, *Geo-
trichum*, and Sterile dark and hyaline forms [108, 111, 114, 122-125].
Basidiomycete fruiting bodies are not abundant on actively forming
acid peats, but they become common if for any reason the peat surface
dries out [126-128]. Mastigomycotina are not uncommon in ombrophilous
peats [129], but they do not appear to form a dominant component of
the mycoflora. Several of the fungi found in acid peat lands are apparently restricted to such habitats, being rare or absent from mineral soils [130].

   Stenton [131] has shown that, by contrast to acid bogs, rheophilous peat has a more luxuriant fungal flora in terms of the variety
of species encountered. Several of the previously mentioned fungi
occur in his lists for English fen peat, and our own pertaining to
the Everglades, but in neither instance are the floras dominated by
species characteristic of ombrophilous peats. *Fusarium*, *Cylindro-
carpon*, *Arthrinium*, *Volutella*, and *Pseudeurotium* are among other
genera common in one or more fen peat types. Zimenko [114] has recorded several species of Mucorales in various lowland bogs.

   Relatively few estimates have been made of the algal populations
at the surface of peats. Timonin [132] recorded about 10,000 algae

per gram dry weight in a valley peat, and Visser [112] has estimated the surface population of tropical *Papyrus* to be of the order of 1,000,000 individuals per gram dry weight. Clearly in some instances algae may be an important component of the surface flora. In the central part of the Everglades slough, where peat accumulates, filamentous blue-green algae, mostly desmids, are very abundant [133]. In the margins of the slough, where the limestone bedrock is close to the surface and Ca more readily available, peat does not accumulate but there are dense mats of carbonate-precipitating organisms [134]. Likewise, in some ombrophilous bogs, especially if partial drainage has caused a breakdown in the closed nature of the surface vegetation, there may be a surface weft of filamentous species, which are mainly *Conjugales*, and in the bog waters an extensive population of diatoms and/or desmids is not uncommon [135-138]. The presence of members of the Cyanophyta on or in peats is important, because of their ability to fix nitrogen and to contribute to nutrition [115].

Animals form an important component of any soil population. A series of accounts has been given of specific groups of animals occurring in blanket bog [111], in alpine peats [139], in *Sphagnum* bogs [140-142], in various New Zealand peats [142], and in English fen peat [143]. However, the methods currently available limit the value of the data that may be obtained from such studies in relation to questions concerning the relative activity of specific groups of animals.

D. Microbiological Activity below the Peat Surface

One must be impressed by the obvious lack of extensive microbial activity within peat that is maintained in a relatively natural state. The excellent preservation of various plant tissues is one measure of this inactivity and an even more dramatic demonstration was the discovery of extremely well-preserved human corpses in peat graves in Scandinavia [144].

However, we are here concerned with the possibilities of more limited or more specific microbial activity that will perhaps result in only relatively small changes over long periods of time. That such activity occurs is suggested by the well-established existence of viable populations of microorganisms at considerable depths in peat bogs. Waksman and Purvis [145] demonstrated the presence of bacteria in all strata of both ombrophilous and rheophilous peats, and since then numerous similar results have been reported. Visser [112] has studied one of the deepest peat deposits known and has found a total of 190 organisms per gram dry weight some 10 m below the surface. In our studies of the Everglades peats we have obtained counts of 1800 bacterial cells per gram dry weight at 3.2 m in a mangrove peat, and 11,000-17,000 per gram dry weight at 1.8 m below a freshwater *Mariscus* community. No fungi grew in cultures from the deep mangrove peat, and only 112-143 per gram were recorded from below the *Mariscus*.

The possibility that these organisms are active at such depths is hard to prove or disprove using existing microbiological methods. Using the washing box technique, Dickinson and Dooley [3] have shown that very few particles from newly exposed deep rheophilous peat from beneath an ombrophilous bog were colonized by fungi or actinomycetes. Using a similar technique, subsurface samples of Everglades peat from depths of 3-4 m were shown to be almost completely free from actively growing filamentous fungi [146] (Table 2). Actinomycetes were quite common on fragments from the surface layers of freshwater, *Mariscus* peat but they were very rarely isolated from deeper horizons or from the surface of the other two types of peat. This technique, however, does not provide any information about the bacteria, which are numerically the most abundant group in deep peat.

There do not appear to have been any systematic direct examinations of deep peats such as might provide evidence of recent microbial activity. Some brief observations have been made by us [146] on Everglades peat using the technique described by Casida [147] in which peat is viewed by incident light using a planapochromat x100 oil immersion objective on an inverted microscope. We observed extensive networks of very narrow mycelium, possibly formed by actinomycetes, in freshly collected peat from depth of several meters; this

TABLE 2

Colonization (%) of Fragments Taken from Peat Cores

and Washed Free of All Detachable Propagules[a]

	Rhizophora (Cores av. 336 cm deep)			Juncus (Cores av. 230 cm deep)			Mariscus (Cores av. 58 cm deep)		
	Fungi	Actinomycetes	Sterile pieces	Fungi	Actinomycetes	Sterile pieces	Fungi	Actinomycetes	Sterile pieces
Surface (2-3 cm deep)	20	5	76	27	2	72	48	55	22
Midpoint of core (see average depth in each instance)	0	0	100	1	0	99	9	6	85
Bottom of core (2-3 cm from peat-mineral horizon interface)	0	0	100	0	0	100	6	4	90

[a]287-300 pieces incubated on soil extract AGAR, pH 7.0, at 20°C for 108 days.
[b]Data from Dickinson et al. [146].

does not necessarily imply aerobic conditions generally at these
levels, because some actinomycetes are microaerophilic. Wider mycelium, such as is produced by many fungi, and bacterial colonies, were
uncommon although individual bacterial cells were very numerous in
some samples in the water between peat particles. However, it is at
present extremely difficult to distinguish between live and dead cells,
using this technique.

At present we are therefore still heavily dependent on indirect
methods for assessing microbial activity. Two approaches are possible,
involving on the one hand, studies of materials deep in peat to determine any changes in chemistry that might indicate biological activity
(this will be dealt with later), and, on the other, a study of the
physiology of various species that are present in deep peats. By comparing their behavior under conditions related to those in the peat,
it is at least possible to determine whether there are major obstacles
to their growth in the natural environment.

Clymo [104] has demonstrated that decomposition of *Sphagnum* takes
place down to 75 cm below the surface. The main environmental factors
examined in relation to growth of organisms in many peat forming systems are temperature, especially the possibility of growth at low
temperatures [148-150], and oxygen tension, which has generally been
shown or assumed to be very low, especially in the deeper layers of
peat.

Dehydrogenase enzyme activity has been measured in ten Everglades
peat profiles by Given et al. [29], using the triphenyltetrazolium
chloride method [151]. High activities were found in the top 50 cm
(0.2-1.6 mgm TPF per gram, wet basis). In most freshwater peats the
activity fell essentially to zero below this level, but in saline
peats modest activity (0.1-0.2 mg TPF per gram) was found throughout
the lower part of the profile. It was suggested that anaerobic organisms might be active in the lower levels of the saline peats because
of the availability of ions from sea water, particularly sulfate. The
observed tendency of pyrite and organic sulfur to increase in concentration with depth is in accord with this suggestion; the sulfur contents of freshwater peats are very low.

Catalase activity and hydrolytic activity, at depths from 0 to 100 cm have been measured in forest peats that contained also some mineral horizons [152]. Both parameters were high in the peat but low in the mineral horizons. Catalase activity in the peat increased in the summer months but was reduced by drainage.

An interesting approach to the potential activity of microorganisms in peat was taken by Kaleja [153]. He fermented peat anaerobically at 50°C in the presence of various media. The ferment was inoculated with a mixed culture of methane-producing bacteria, but otherwise the natural microflora was relied on. After 28 days, the populations of cellulose and other carbohydrate-decomposing bacteria, of denitrifiers, and of ammonifiers had increased greatly, whereas the small initial populations of sulfate reducers and methane producers had actually decreased.

The possibility of chemical inhibitors of microbial activity in peats is occasionally raised. The evidence has been reviewed by Küster [154]. On balance it appears that peat extracts stimulate the growth of fungal mycelia, although this effect was greater if the extract was filtered through charcoal. A water eluate of the used charcoal indeed had an inhibitory effect. This aspect of the microbiology of peat obviously needs further investigation, but the indications that there are substances present that can both stimulate and inhibit microbial growth are interesting, because of the implication of possible natural control mechanisms.

E. Summary of Microbiological Aspects

As with other aspects of peat studies, it is essential for microbiologists to define the peat type being studied using accepted ecological criteria and terminology. Far too many published studies cannot be used in any detail because of deficiencies in this respect. Likewise, the developmental status of the peat system is critical in terms of microbiological studies, and it is clearly important to know whether the peat is still forming, whether it is preserved in a seminatural state, or whether it and its associated angiosperm community

are being destroyed by any agency. Microbiologists should also recognize that the physical nature of peats may demand their own methods
for study and certainly present problems when it comes to expressing
data in a meaningful way. It is hard to visualize a gram of dried
peat, and yet it is also difficult to compare a cubic centimeter of
peat with a gram of mineral soil.

VI. ANALYTICAL BIOCHEMISTRY OF PEATS

A. Introductory Remarks

The customary means of obtaining a chemical characterization of
peat is to carry out a proximate analysis, in which various fractions
are extracted with appropriate solvents or removed by hydrolysis.
These operations may be carried out sequentially on the same sample
[113, 155], or on separate samples [156]. The objective is to dispose
of 100% of the sample into boxes variously labeled "humic acids,"
"lipids," etc. The difficulty comes in deciding on the labeling of
the boxes. The analytical schemes were originally based largely on
methods used for fractionating the constituents of tissue of fresh
plants, with the addition of a humic acid extraction. It was assumed,
for example, that material removed from peat by the reaction conditions used to remove hemicellulose from fresh wood is properly described as hemicellulose. Colorimetric methods for determining total
sugars or reducing sugars have commonly been used for analyzing hydrolysates, although these, of course, permit no qualitative identi-
fication of the original polymer (or mixture of polymers), and in any
case are now known to be unreliable for hydrolyzates.

The common custom in the recent Russian literature is to play
safe by having two categories described as "easily hydrolyzed carbohydrates" and "difficultly hydrolyzed carbohydrates." However, the
"hemicellulose" content of the first and the "cellulose" content of
the second are usually also reported. It is difficult to envisage any
simple but reliable method of determining these plant polysaccharides

in mixtures if one admits that microbial polymers may be present as well and may contain at least some of the same sugar monomers as cellulose and the polymers making up "hemicellulose."

Lucas [157] has observed that the loss in weight when humins from peat are hydrolyzed by acid is much greater than the weight of sugars he could isolate and identify by ion-exchange purification and gas chromatographic analysis. It is well known [158, 159] that in the acid hydrolysis of polysaccharides some part of the sugars is converted to furfural derivatives, and some unhydrolyzed dimers always remain. Moreover, the loss in weight will include some amino acids that had been bound to the carbohydrate polymers [160]. These effects, however, are not sufficient to explain the very large discrepancy. Therefore, it is quite likely that some plant lignin has been partially degraded to the point where it may be solubilized by hot acid.

With the increasing use of powerful chromatographic techniques, the emphasis in peat studies, as in other fields, has shifted in the direction of determining individual compounds of proven structure. The price paid for basing analyses on well-defined compounds is, of course, that one can probably identify only a small fraction by weight of the whole peat. Moreover, it is usually desirable to carry out a preliminary fractionation of the sample, and one is then back with the problems of the "proximate analysis" approach.

In what follows the composition of peats in terms of proximate analyses will be discussed briefly, and then a more extended discussion of the main structural types of compound will be presented. The older literature on peat biochemistry was reviewed by Francis [34] and some more recent work was broadly discussed by Swain [16].

B. Composition of Peats in Terms of Proximate Analyses

Because different workers have used different fractionation schemes, comparisons are not always possible. However, there seem to be a few established generalizations. The humic acid contents of rheophilous monocotyledonous and forest peats are generally high (30-80%), whereas those of ombrophilous *Sphagnum* peats are typically

only 5-25% [30]. Yet the sedge, reed, and mangrove forest peats of
Florida are of rheophilous type and have low humic acid contents
(5-20%). There appears to be some connection between the degree of
decomposition, measured microscopically or by sieve analysis, and
the humic acid contents, at least in European peats from temperate
zones [24, 24]. There is often a trend of increasing humic acid content with increasing depth [16, 24, 155], although the trend is by
no means continuous. Indeed, Bugubaev [16] found the humic acid contents of peats at four sites to be highest in the middle of the profile. In Florida peats, the variation with depth is largely random.

TABLE 3

Composition of the Upper Layers of the "Galitskii Moss"

Peat Deposits (High Moor Peat Bed)[a]

Depth (cm)	Degree of decomposition (Botanical analysis) (%)	Organic Substance of Peat (%)				
		Bitumen	Humic acids	Lignin and cutin	Hemicel-lulose	Cellulose
0-5	0	2.4	1.9	9.2	36.2	21.6
5-24	5	3.5	5.0	10.1	16.6	29.6
24-31	26	7.1	25.4	17.3	13.0	22.6
31-39	15	4.2	15.1	10.5	25.5	21.7
39-44	32	7.4	26.4	21.3	17.2	16.3
44-50	14	4.1	15.4	19.7	15.7	23.8

[a]Data from Kurbatov [24].

Kurbatov [24] observed that high moor peats often contain interlayers of more highly decomposed peat. He shows proximate analyses
(Table 3) for a 50-cm section, from which it can be seen that both
the degree of decomposition and the humic acid contents are highest
in the 24 to 31 and 39 to 44-cm levels. Kurbatov suggested that most
of the essential phenomena of peat formation occur aerobically in
whatever is the surface layer at any given time, and once this layer
is buried beneath fresh peat little further change occurs. He suggests

that in periods of warmer and drier summers, when the water table is lower, more intense aerobic alteration of the plant debris can take place.

The idea that the character of peat at any level in a profile represents a memory of prevailing conditions when that level constituted the surface seems a very reasonable one for any type of peat.

In subtropical and tropical climates, it is common to have hot wet summers and dry but warm winters; the freshwater peats of southern Florida often dry out in the winter, and within recent memory this has been much more severe in some years than others. It appears, therefore, that we should expect to find nonlinear relations between humic acid content or degree of decomposition and depth, and other chemical characteristics may well vary nonlinearly also.

Too few peats from tropical and subtropical areas have been studied for it to be clear whether they differ characteristically in proximate analysis. In their low humic acid contents and high fulvic/humic ratios the rheophilous peats of south Florida [157, 160, 162] and of Selangor, Malaysia [163], are anomalous, in that the values resemble those in highmoor *Sphagnum* peats [25, 48]. On the other hand, some peats from Ceylon are reported to have very high humic acid contents and less cellulose than north Russian rheophilous bogs of comparable degree of decomposition [25]. It appears that some generalizations about the composition of the various types of peat in temperate zones are reasonably valid, but they may not hold for peats in warmer climates.

That part of a peat soluble in organic solvents[*] usually represents a fairly minor constituent (2-10%), although higher values have sometimes been reported. Rakovskii et al. [25] believe that peats derived from small shrubs have high contents of organic-soluble material, and attribute this to the high content of such material in

---

[*]This fraction of peat is often termed "bitumen." We strongly object to the use of this word in this context, because it has already been preempted by petroleum chemists and geochemists for material of quite different structure and origin. To describe this fraction as lipids is also objectionable, since it contains a great deal of non-lipid material, such as resins and phenols.

the original plants. Proximate analyses shown by some authors contain entries for pectin and cutin. The evidence for the presence of pectin in peats presumably rests mostly on the occurrence of uronic acids in hydrolysates, but this evidence by itself is not acceptable, because bacterial slimes and capsules often contain polyuronic acids. Naucke [164] refers to an unpublished thesis [165] that appears to have described intact polymeric pectins isolated from peats, and to have given molecular weights for the polymers (20,000-70,000, and 170,000 for *Sphagnum* peat). The pectin content of peats is said to be 2.5-13.7% (166).

C. Humic Acids and Phenolic Substances

No attempt will be made here to review in any comprehensive sense the biochemistry of humic acids, particularly since, on the evidence so far available, there is little reason to suggest that the humic acids of peats differ in important ways from those in mineral soils. Peat humic acids have been reviewed by many authors [16, 24, 25, 30, 34, 167, 168].

The C:H ratios of fulvic acids in peats are high, about 1.85, compared with values of 0.60-0.65 for soils, indicating the peat fulvic acids to be more highly substituted and more aromatic. The contents of carboxyl and hydroxyl groups were relatively high in the acids from peat [170].

Various authors have obtained molecular weight distributions for humic acids by gel permeation chromatography (e.g., 169). The reliability of the method for humic acids is questionable, because the "molecular weights" or micelle weights of any preparation vary with pH and the nature and amounts of any adsorbed cations, as has been shown recently using the equilibrium ultracentrifuge method [57, 171]. Humic acids from lignites and from two peats from a bog near Lake Balaton, Hungary, were rigorously purified from cations and studied by this method. Micelle weights for the peat humic acids changed less with pH than those from lignites; for peat humic acids a value of 4400 at pH 5 decreased to 3000 at pH 6.5. At higher pH

the micelles disaggregated and became unmeasurable, so that the true
molecular weight was quite low. Micelle weights increased in solutions
on standing and on adding various ions (Cu^{2+}, Fe^{3+}, Co^{2+}, Mg^{2+}) in
concentrations in the range 1-10 mmole/liter.

It is well known that the oxidative degradation of soil humic
acids yields a mixture of phenols among the products [172, 173].
More recently humic acids from four peats of different degrees of decomposition have been degraded by alkaline nitrobenzene oxidation
[174] and eight phenols were identified in all samples. As when soil
humic acids are similarly treated, the phenols bear obvious structure
relationships to lignin, e.g., syringaldehyde (1), vanillin (2), and
p-hydroxybenzaldehyde (3). The lignin structure is built of phenylpropane units (4), in which the positions marked 3, 4, and 5 are substituted by oxygen groups as in (1), (2), and (3). The syringyl type
(1) is largely restricted to the lignin of monocotyledonous Angiosperms.

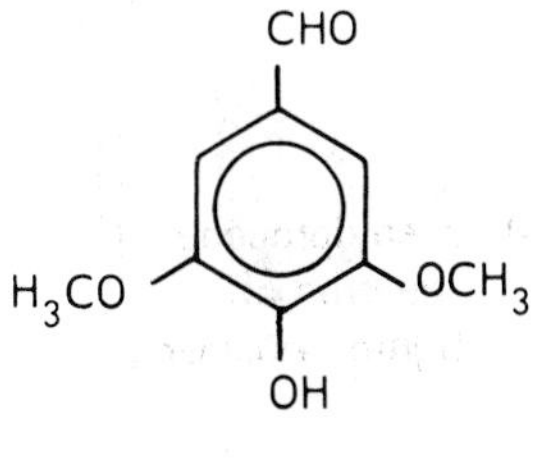

1. syringaldehyde

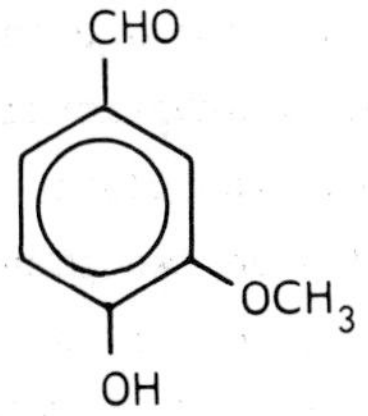

2. vanillin

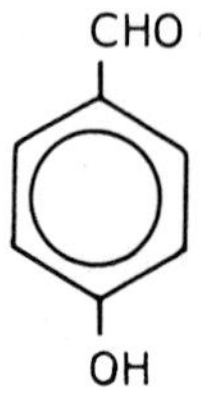

CHO

OH

3. p.hydroxy—
benzaldehyde

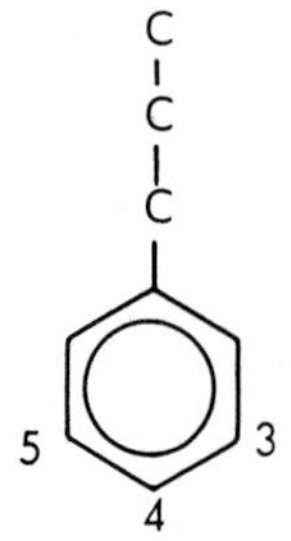

C
|
C
|
C

5   3
4

4. phenylpropane
skeleton of
lignin monomer

Burges et al. [175] have successfully used reductive degradation
of soil humic acids with soldium amalgam, and obtained a number of
phenols and phenolic acids. Many of these bore clear structural
resemblances to lignin, but some had the 1,3-dihydroxy- or 1,3,5,-
trihydroxybenzene pattern of substitution (5) and (6), and these, it
was suggested, must derive from flavonoid structures, such as (7).
Splitting of ring C would give two phenols, that from ring A being
of the catechol or lignin type (2), and that from ring B a resorcinol
or phloroglucinol type, e.g., (8). This line of reasoning therefore
implicates the flavonols of higher plants as well as lignin in
humic acid formation.

5. resorcinol

6. phloroglucinol

7. quercitin, a flavonoid

8. scission product from
B ring of flavonoid

Martin and Haider [169] observed that certain molds, including the soil organism *Epicoccum nigrum,* deposited a number of phenols in their growth medium. Among these were several that had the resorcinol or phloroglucinol structural pattern. The phenols condensed with some amino acids also secreted by the mold to form a product closely similar to the humic acids of soils. Reductive degradation yielded again the same range of phenols. It is possible, therefore, that the phenols noted in degradation by Burges et al. [175] were partly or wholly derived from mold metabolites.

The reductive degradation approach, using sodium in pyridine, has been applied to fulvic acids from peats by Wildenhain and Henseke [176]. The fulvic acids were isolated by means of ion-exchange resins. The phenols found were 2,6-dimethoxyphenol (*9*), *o*-methoxy-catechol (*10*), and diguaiacylmethane (*11*). In (*9*) and (*10*), the propane side chain has been completely removed, but the presence of the methoxy groups indicates a lignin origin.

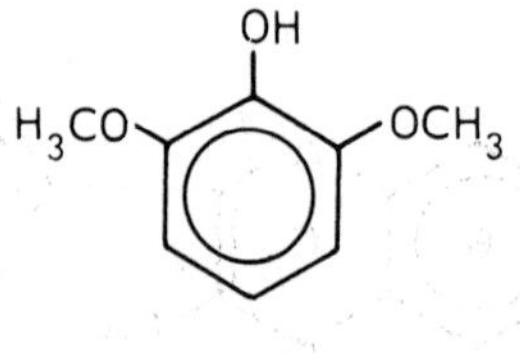

9. 2,6–dimethoxy phenol

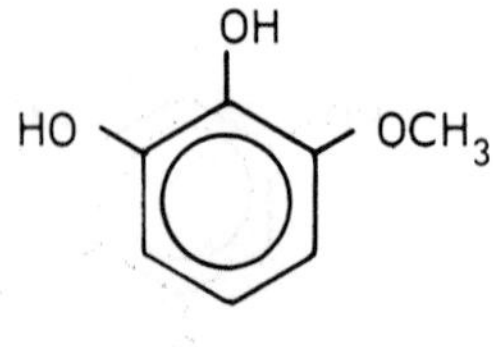

10. o.methoxycatechol

Some authors have found free phenols in peats, simply extractable without chemical degradation. In this way, various phenol carboxylic acids were extracted [177] from two peats having degrees of decomposition 30-33% and 55-60%, respectively; the latter had the lower content of acids. The acids identified were all of the lignin type: *cis*- and *trans*-ferulic (*12*), *p*-coumaric (*13*), syringic (*14*), vanillic (*15*), and *p*-hydroxybenzoic (*16*). Measurements were made in May, June, and September and the highest concentrations were found in May. It has been suggested that phenols derived from lignin degradation are instrumental in trapping and accumulating U, Ge, and V in peats [178].

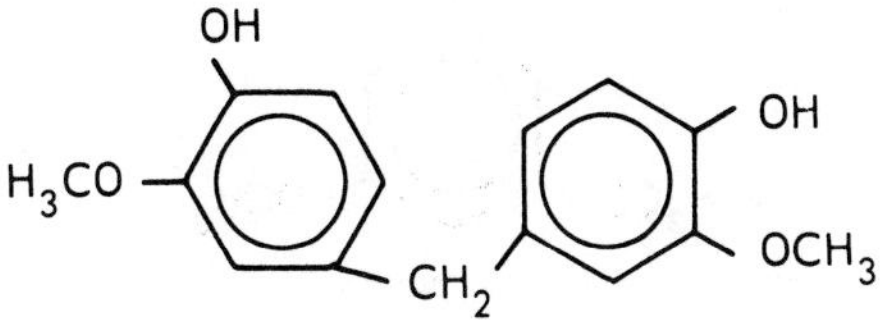

II. diguaiacylmethane

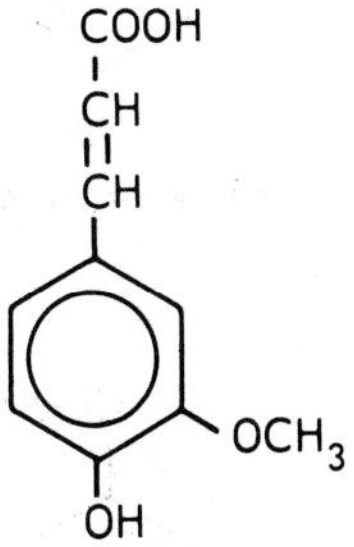

I2. ferulic acid

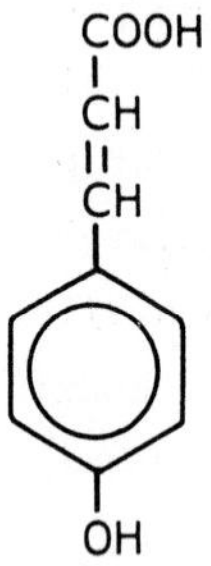

COOH
|
CH
||
CH

OH

13. p.coumaric acid

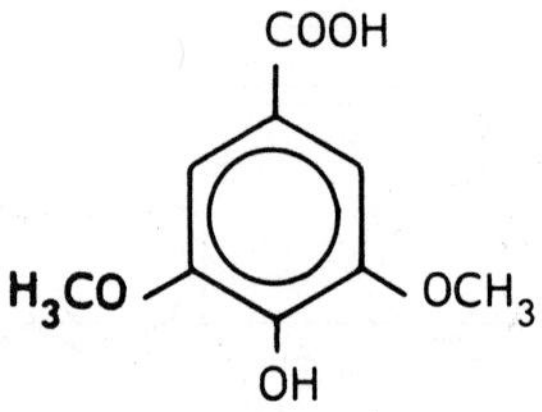

COOH

H₃CO   OCH₃
OH

14. syringic acid

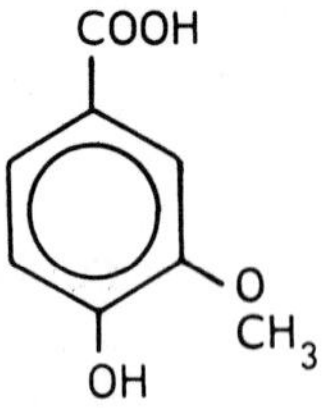

COOH

O
CH₃
OH

15. vanillic acid

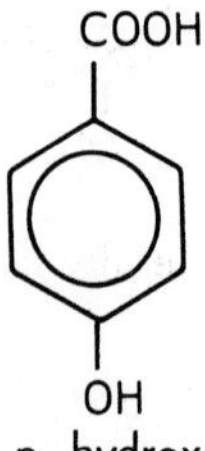

COOH

OH
16. p. hydroxy–
     benzoic acid

The vascular plants contain a wide variety of phenolic substances, usually as glycosides, but the phenols of the mosses have been little studied; members of the Bryophyta contain little or no lignin. However, Bendz et al. [179] have found some flavonoid substances in the red moss, *Bryum*. Morita [180] has analyzed phenols found in acid and alkaline hydrolysates of whole *Sphagnum* plants, and found p-coumaric acid (*13*) (the most abundant component), ferulic acid (*14*), protocatechuic acid (*17*), α-resorcylic acid (*18*), resorcinol (*5*), phloroglucinol (*6*), orcinol (*19*), and acetovanillone (*20*). The hydroxycinnamic acids (*12*) and (*13*) occur as such in vascular plants as biosynthetic precursors of lignin and are found among the microbial degradation products of lignin [181]. Compounds (*17*) and (*20*) can be formed from these by oxidation. Thus *Sphagnum* evidently contains phenolic structures closely related to those in vascular plants. The remaining compounds (*5*), (*6*), (*18*), and (*19*) have a different substitution pattern and could be precursors or breakdown products of flavonoids. Presumably when *Sphagnum* forms peat, all of these types could be released and be available for the build-up of humic acids.

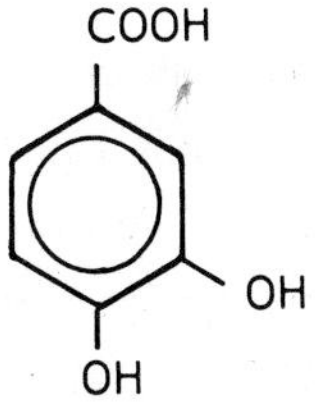

COOH

OH

OH

17. protocatechuic acid

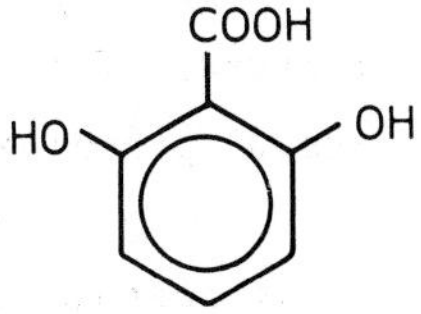

COOH

HO   OH

18. α—resorcylic acid

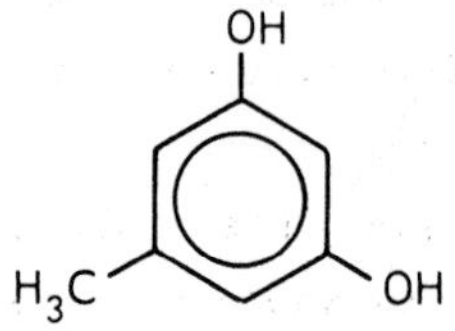

19 orcinol

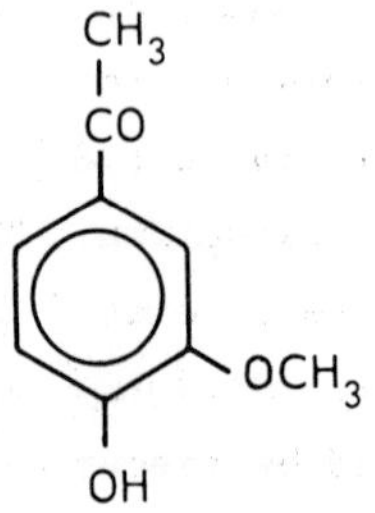

20 acetovanillone

The phenols extractable from a *Sphagnum* peat include (*2*), (*3*), the glucoside of (*20*), (*12*) and its 3-methoxy derivative (caffeic acid), (*14*)-(*17*), and various compounds structurally related to (*12*) and (*13*), such 3,4-dihydroxypropiophenone (*21*), and *p*-hydroxylphenyl pyruvic acid (*22*) [182, 183]. Other phenols included 2,4,6-trihydroxy-acetophenone (*23*). Thus, in the peat as well as in the fresh plant we have a range of lignin-related and flavonoid-related phenols. It is interesting that in several of the peat analyses noted above methoxy groups survive in a number of compounds because in laboratory stimulation of natural microbial processes, these groups are rapidly demethylated [181].

Sphagnum contains up to 1% acetone-soluble phenols, in addition to phenol carboxylic acids [184, 185]. *Sphagnum* is said to contain a substance known as sphagnol, a term apparently first used by Czapek [186] to describe the lignin-like substances of mosses. Dark-colored

21. 3,4–dihydroxy
propiophenone

22. p.hydroxyphenyl–
pyruvic acid

23. 2,4,6 trihydroxy–
acetophenone

material similar to natural humus is not formed in laboratory experiments on microbial decay if the sphagnol is first removed [24]. The
structure of sphagnol has not been reported; one can infer from the
literature that it is thought of as insoluble unless first chemically
degraded by, for example, alcoholic HCl. On the other hand, *Sphagnum*
is said to contain antiseptic substances [25] and the peat derived fro[m]
it has long been reputed to have antiseptic properties. These properties are adduced in support of the view that *Sphagnum* contains
specific inhibitors that restrict microbial decay and contribute to
the preservation of the peat (e.g., [25]).

The well-known work of Flaig [181, 187] shows how microbial degradation products of lignin can be further altered by extracellular
enzymes and finally condensed into humic acid-like products. He also
shows [187] how amino acids may combine reversibly or irreversibly
with the intermediates in this process, and thus become incorporated
in the humic structure. There seems no reason why in a natural
system other phenols, such as mold metabolites, 0-diphenol oxidase
degradation products of flavonoids, or compounds derived from mosses,
should not be involved in essentially the same processes [185].

Mathur and Paul [188] found that species of the genera *Penicillium*, *Aspergillus*, *Arthrobacter*, and *Pseudomonas* were capable of
utilizing purified "mobile" humic acids as a source of carbon and/or
nitrogen. The humic acids were derived from the Melfort Orthic Black
soil. In experiments with a *Penicillium* isolate, the surviving humate
was found to have increased in contents of nitrogen and hydroxyl
groups. Under conditions of restricted aeration, the soluble degradation products were found to contain saligenin (*24a*) and salicylaldehyde (*24b*). Saligenin is of restricted occurrence and is found as
salicinin, the 1-glucoside, in the leaves and bark of most species
of *Salix*, whereas the benzoyl derivative is found in leaves and bark
of *Populus* spp., which are also members of the Salicaceae. Willows
and poplars were the main elements of the population, in addition to
grasses, inhabiting the site where the soil was obtained before it
was cleared for cultivation in about 1900. Members of the Salicaceae,
incidentally, are common in peat-forming environments.

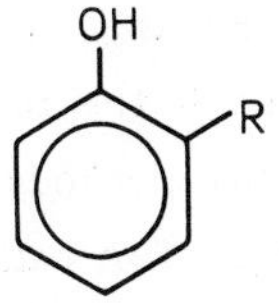

24.
a. saligenin
R=CH$_2$OH
b. salicylaldehyde
R=CHO

Saligenin is known to undergo condensation to resin-like substances and to have a great affinity for amino acids. Mathur and Paul therefore point out that it could be involved in soil humus formation. This raises two interesting issues. First, we apparently have a taxonomic memory in the structure of the humic acids, and second, because saligenin is not related to lignin or flavonoids, the data show that other phenols from the higher plants can contribute to humic acid formation.

As already noted, vascular plants contain a wide variety of phenolic substances, other than lignin and flavonoids. Some of these, like hydroxystilbenes and gallic acid derivatives, are involved in the structure of the nonflavonoid tannins. Others, like the cinnamic acid derivatives, presumably are ubiquitous because of their role in the biosynthesis of phenylalanine, lignin flavonoids, and other substances. The hydroxytropolones found in trace amount in the heartwood of trees of the Taxodiaceae (cypresses and cedars) are strongly fungicidal and evidently exercise a protective function, which may well continue after the death of the organism or the shedding of its woody parts. Other structural types, like the hydroxyxanthones, hydroxycoumarins, and lignans, have no clearly defined function. Rotenone, from the root of *Derris*, is a hydroxyxanthone and is used in insecticides. Several useful reviews of phenolic substances in plants have been published [189-193]. Some of these structural types exhibit a fair degree of taxonomic specificity. Which of them are likely to

survive the death of the plant and be capable of incorporation into
humic acid structure is not known, though this is obviously a matter
worth exploring, employing both the microbial approach of Mathur and
Paul [188] and purely chemical methods.

The use of plant extracts for tanning hides has been known for
thousands of years, and tannins have been the subject of scientific
study for over 100 years. Yet it is only in the last few years that
a reasonably clear picture of the nature and distribution of the
condensed tannins has emerged. This clarification has hardly escaped
yet from the specialist literature, so that there are statements in
the peat literature that are based on misconceptions.

It is conventional to distinguish between the hydrolyzable tannins
which are glucosides of gallic acid (25) and its derivatives, and the
condensed tannins, which are low molecular weight polymers of certain
compounds of the flavonoid class. The gallic acid derivatives no
doubt enter peat but are of no great interest for present purposes.
The major and accessory pigments of most flowers are flavonoids (an-
thocyanins; flavones; flavonols, e.g. (7); aurones). They occur also

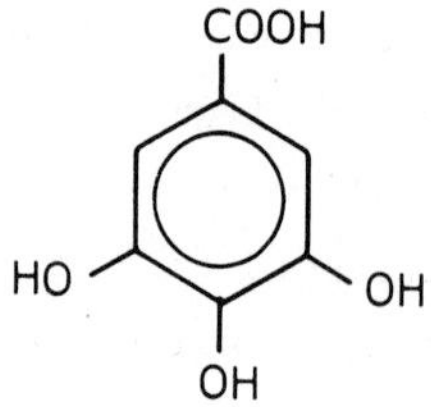

25. gallic acid

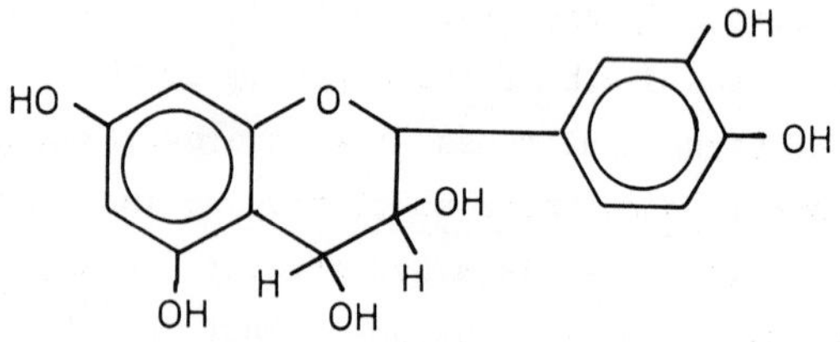

26. leucocyanidin

in leaves and fruits, and to some extent in other organs; their
weight concentration is not high. The condensed tannins usually have
as their monomer units flavan-3,4-diols or leucoanthocyanidins (e.g.,
(26), although in a few plants catechins (same structure, but lacking
the 4-OH group) are minor or major constituents of the polymers.

TABLE 4

Tannin Contents of Mangroves

Family	Genus and species	Tissue	Tissue, %, dry basis Tannins	Nontannins
Rhizophoraceae	*Rhizophora candelaria*	bark	25-30	--
	R. mangle L	wood		--
		bark	20-30[a]	6-11
	R. mucronata	bark	29-40	--
	Bruguiera gymnorrhiza	bark	43	--
Combretaceae	*Laguncularia racemosa*	wood	5	2-7
		bark	12-24	7-17
		leaf	30	14
	Conocarpus erectus	wood	8	1-5
		bark	12-22	6
Verbenaceae	*Avicennia germinans* (=*nitida*)	wood	ca. 0	--
		bark	ca. 0	--
	A. officinalis	bark	5	--

[a]Russell [196] states that *R. mangle* wood has "insignificant
tannin content." Chen [193] found about 5% identified leucoanthocyanidins; a crude tannin extract would amount to considerably more
than this.

Leucoanthocyanidins are of extremely wide occurrence in vascular
plants [190, 194], including the pregymnospermous Cryptogams, although
in the Angiosperms they are less common in the monocotyledons than
in the woody dicotyledons. Usually they represent a fairly small
fraction of total organic matter, but in organs of some woody plants,
they may represent 20-40% of total organic matter [190, 195]. The
genera in whose species high concentrations of condensed tannins
occur include *Acacia* (of the Leguminosae), *Uncaria* (Rubiaceae),

Schinopsis, *Rhus*, and *Astronium* (all of the Anacardiaceae), *Rhizo-phora* (Rhizophoraceae), *Pinus*, *Tsuga*, *Picea*, and *Larix* (all of the Pinaceae) and *Eucalyptus* (Myrtaceae). Of these, a number of members of the Pinaceae and many of the mangroves are known to give rise to peat in suitable environments in many parts of the world. Indeed, the Rhizophoraceae include four other tannin-rich genera, and the Combretaceae and Meliaceae also contain such genera. The only man-grove that contains little tannin is *Avicennia*, of the Verbenaceae. Table 4 shows the tannin contents of some mangroves, determined as the fraction of extractable material that is removed from solution by a standard hide sample. The nontannins contain a variety of sub-stances, but chiefly phenols of low molecular weight that do not pre-cipitate on hide powder. The data are taken from Hillis [190] and Russell [196]. It can be seen that polyphenolic substances account for very substantial proportions of the organic matter of some tissues of these peat-forming plants.

The condensed tannins, as already stated, are polymers of flavan-3,4-diols, e.g., (*26*). Polymerization involves a loss of the elements of water from the OH group in the 4-position and the hydrogen in, most commonly, either the 8 or 6 position [see (*27*)]. Oligomers

27. a dimeric unit within a
leucoanthocyanidin polymer

(2-8 monomer units) are soluble. In mature or old tissue (heartwood, bark) they polymerize further so that they become insoluble in water, although they will be still partly soluble in hot alcohol, acetone, or ethyl acetate. The oligomers rapidly polymerize further in aquous acid, although in propanol-HCl or butanol-HCl, polymers are broken down to monomers and converted to highly colored anthocyanidin salts, a characteristic used in their analysis [190].

The extraction of optically dense cell fillings has been followed microscopically for a number of species [190]. It is often found that only comparatively small amounts (4-15%) are extracted by a succession of solvents (hot water, methanol, acetone, etc.) and the residual tissue still contains abundant cell fillings. If the tissue is instead extracted with hot 0.5% sodium hydroxide, the extract is 20-40% of dry organic weight and the cell fillings are then no longer visible. The material extracted with alkali, however, contains carboxyl groups and appears to be a mixture or copolymer of somewhat oxidized leucoanthocyanidins with a small amount of lignin [197-199]. The pyran ring of flavonoids is easily ruptured in alkaline media, and phenols are particularly susceptible to oxidation at high pH. It seems possible that similar changes of leucoanthocyanidins could occur at nearly neutral pH when tissue decays and cell contents become exposed to enzymatic or aerial oxidation. We see from the above that some plants that give rise to peat contain tannins as important constituents of many of their organs. Little study has been made, however, of their fate in peat accumulation.

Chen [193] (see also [95]) identified the principal leucoanthocyanidins in the leaf, stem and roots of the red mangrove *R. mangle*. In the woody parts, polyphenols precipitated by lead acetate represented about 12% of total organic matter, the leucoanthocyanidins accounting for about one-half of this. As already noted, dense cell fillings are abundantly seen *in situ* in decayed parts of *R. mangle* in surface litter and in peat, yet Chen was unable to detect leucoanthocyanidins in mangrove peats, even when vigorous extraction conditions were used. She was able to detect small amounts in senescent

yellow leaves, but none in decayed blackened leaves. Seen under the
microscope, senescent leaf tissue still contained a large number of
dense cell inclusions after extraction with hot methanol. However,
if the tissue was extracted at room temperature with dilute sodium
hydroxide, the cell fillings largely disappeared. Moreover, the
extract was precipitated by dilute acid and had an infrared spectrum
very like that of the peat humic acids. It appears, therefore, from
the optical evidence that tannins survive the death of plant organs
and remain to be commonly found in tissue of the peat; but they have
been altered (further polymerized and oxidized?) in such a way that
they are no longer soluble in organic solvents or hydrolyzable by
alcoholic HCl and have become humic acid-like.

The presumed flavonoid phenols found in the degradation products
of humic acids by Burges et al [175] need not be derived only from
leucoanthocyanidins, because vascular plants contain other flavonoids
as well. Chen found quercetin (7) in moderate abundance in the leaf
of *R. mangle* but detected none in decayed leaf or peat. All classes
of flavonoid appear susceptible to change during senescence [200].

There is another aspect of tannin biochemistry that may be relevant to peat. Tannins have the power of combining not only with
the protein of skin but with gelatin in solution, with the enzymes
of saliva, and with other enzymes. It has therefore been suggested
that they play an important role in the defense of higher plants
against microbial attack; Harborne [201] admits the possibility but
regards it as still unproved. It has also been suggested that tannins exert a protective action *post mortem*. It is common practice
in the maritime parts of the far east for fishermen to steep their
nets in tannin extracts in order to preserve them; evidently the
tannins deactivate cellulolytic enzymes [190]. Cohen's association
of the preservation of aerial organs of *R. mangle* but not *A. germinans*
with the presence of tannin inclusions in the former has already
been noted [1]. The possibility of *post mortem* protection of plant
parts against decay warrants further study, as does the role of
flavonoids in humic acid formation.

Some recent work that contributed greatly to our understanding of tannin structures suggests a possible new approach to the fate of these substances in litter and peat. Thioglycolic acid efficently cleaves the C-C bond linking the flavan molecules in such dimers as (27) [202-204]. The tannins extracted from the bark of the western hemlock, *Tsuga heterophylla*, were depolymerized by this reagent and the monomers identified. If the exhaustively extracted bark was similarly treated a further yield of the same monomers was obtained. Since the reaction was successful with highly polymerized insoluble material in bark, it might also be useful with altered flavonoid polymers in litter and peat.

D. Amino Acids

In the literature, most of the analyses of amino acids in peats were performed by paper chromatography, which has limited resolution and has difficulty in detecting and determining minor components. The concentrations of amino acids extracted by cold dilute acids are small; Kadner et al. [205] report 10 protein acids in a 2% HCl extract of a peat from Germany, and also β-alanine and glucosamine. Rakovskii and Pal'min [206] found that amino acids in acid hydrolyzates account for 5.7-10.3% of the organic mass of the peat and for 28-58% of the total nitrogen content; their studies included peats in moss, grass, and forest communities. Amino acids found in a *Mariscus* and a *Rhizophora* peat in Florida were in a similar range [95, 160, 207]. Nevertheless, Swain reports only 7-10 acids with lower total concentrations (mostly 1-2%) in various Minnesota bogs [38, 208, 209].

In the case of the Rossburg bog, Swain identified -- in the peat -- various tissues of four species of mosses and also of a dicotyledon [210]. He analyzed the amino acid distributions in several fresh specimens of each of these plants, with results shown in Fig.2; it will be seen that there are some differences between different specimens of the same species. Swain concludes that the total amino acid concentrations in the fresh plants are a little higher than those in the surface layers of the peat at the same site (and a little

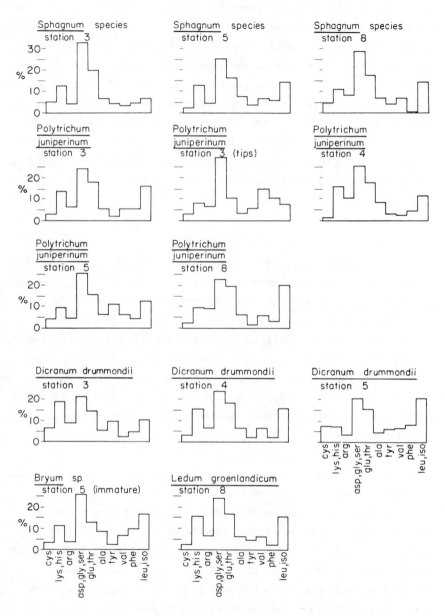

FIG. 2.   Percentage distribution of protien amino acids in plant species from Rossburg peat [38].

lower than those at lower depths). The relative abundance of the various acids was similar, but by no means identical, in plants and sediment.

Amino acid contents are certainly different for different organs of a plant, the content in leaves being relatively high. Nevertheless, it is clear that the concentrations found in peat by Rakovskii and Pal'min [206] and by Casagrande and Given [207] are considerably higher than one would expect to find as an average for a whole plant, presumably as a result of the conversion of plant material plus some input of nitrogen into microbial protein.

In three swamps of different types, 17 common amino acids were found [206, 211, 212]. Some acids (glycine, alanine, glutamic acid, and leucine) are partially released under mild conditions of hydrolysis (1% H_2SO_4), whereas others (valine, lysine, arginine, and proline) are only released under severe conditions (20% HCl for 24 hr at 108-110°C). The amino acid distributions were independent of the degree of decomposition of the peat over the range 5-50%. The fine grained material (humic) contained 1.5 to 2fold more hydrolyzable N than the fibrous components of the peat. The distributions in the peat hydrolyzates differed little from those in the peat-forming plants, and it was concluded that all amino acids in the plants participate in the formation of peat in proportion to their initial content in the plants. In contrast, Kurbatov [24], in a discussion of the genesis of humic acids, says: "The amino acid complexes, it seems, originate mainly from the cytoplasm of dead microorganisms," a conclusion with which we are in agreement.

A hexosamine has been found with amino acids in a Bulgarian peat [213], and the amino acid distributions in a humic acid hydrolyzate were shown to differ from those in whole peat hydrolyzate [214]. Galactosamine is also reported [164]. There appears to be no systematic relation between the amount of any given amino acid and depth within a peat profile [215], but the maximum amounts of all acids are said to be found in the lower layers of the bed [16, 203]. It is suggested that humic acid micelles with attached amino

acids migrate downwards, to be precipitated or filtered out at the
lower depths of the sediment [16].

A comprehensive study of amino acids in peats was made in two
profiles in the Florida Everglades [95, 160, 207, 216]. One site was
saline, inhabited by various mangroves, predominantly *R. mangle;* the
other site was a freshwater environment populated at the present time
by *Mariscus jamaicensis,* a sedge, but was also populated frequently,
during the accumulation of the sediment, by the water lily, *Nymphaea*
sp. Gas chromatographic analyses of volatile derivatives were made
on the free acids (extractable by cold dilate HCl) and on hydrolyzates
of whole peat and fractions of it. Organs of the living plants and
samples of surface litter and decaying root material were also analyzed.

A variety of nonprotein acids was found. Some of these, notably γ-amino-butyric, α, γ-diaminobutyric and α, ε-diaminopimelic
acids, are of frequent occurrence in the mucocomplex of bacterial
cell walls, but are present in low or zero concentration in the higher
plants. Lysine and glutamic acid also occur in the mucocomplex, but
as D forms, rather than the L forms found in proteins, a fact not yet
made use of in geochemical studies. Other amino acids found in the
peats are often associated with the catabolic activities of microorganisms but are by no means specific to them; these include β-alanine, ornithine, citrulline, cysteic acid, and α-aminobutyric acid.
Still other acids (e.g., sarcosine, norvaline, norleucine) are not
found in the higher plants to any extent but are found in soils
[217, 218]; their origin and function are not clear, but they appear
to be associated with microbial activity.

In the *Mariscus/Nymphaea* peat, there was a clear increase in
total amino acid concentration in the sequence fresh plant→litter→
peat (cf. *Mariscus* plant, 18.5 mg/g; litter, 33.4 mg/g; surface layer
of peat, 87.6 mg/g, and the ratio, nonprotein/protein acids x 100,
changed in the same sequence, 2.0→8.4→5.2 (the ratio in the peat was
6.9 at 3 ft and 7.5 at 7 ft). A similar picture was seen at the
saline site, although less clearly defined partly because a layer of
organic-rich lime mud some 50 cm thick overlay the peat. The data
in Table 5 are of interest in relation to this site.

TABLE 5

Amino Acid Concentrations
at Little Shark River Site,
Florida Everglades

Sample hydrolyzed	Concentration of acids mg/g			$\dfrac{\text{Nonprotein}}{\text{Protein}}$ x 100
	Protein	Nonprotein	Total	
Stem of *R. mangle*	13.0	0.39	13.4	3.0
Buried stump of *R. mangle*	21.8	4.7	26.5	21.5
Peat, 6-ft level	42.3	7.8	50.1	18.3
Peat, 12-ft level	29.2	9.9	39.0	33.8
Peat, 13-ft level	2.06	0.93	3.0	68.9

The buried stump (^{14}C date = modern) was evidently the remains of a substantial tree (diameter of stem, 25 cm) that had been blown down in a hurricane. The broken surface was at a depth of some 40 cm. It could hardly retain its original protein, yet it contained slightly more protein acids and much more nonprotein than a fresh stem. Within the peat, the nonprotein/protein ratios are high, as can be seen, and there is a curious increase in the ratio but a decrease in total concentration at the base of the core, near the limestone bed-rock contact.

The characteristic distribution of protein amino acids in the woody parts of *R. mangle* (stem and secondary roots) was to some extent recognizable as a mangrove signature in a rotted stem from the surface, a decayed root, and the 4-ft level in the peat (not in the buried stump, nor in the surface layer of the peat). With *Mariscus* at the other site, the fresh plant signature was more or less recognizable in the decayed leaves but not in the peat.

The bulk of the amino acids in the peat was released on hydrolysis of the humic acid and humin fractions, and the distributions of both protein and nonprotein acids were generally similar in the

hydrolyzates of the two fractions. At most levels, the free (pH-1
soluble) acids represented only 0.1-1% of total acids, although at
the surface level of the *Mariscus* peat, the proportion was 3.1%.
However, it was this fraction that showed the most radically different
protein acid distributions and the highest ratios of nonprotein/pro-
tein (10-40%). The fulvic acid hydrolyzates were also rich in nonprotein acids and the detailed distributions tended to resemble those
in the free fraction.

In fresh *Mariscus* plants, small amounts (chromatographic peak
heights 1-5% of the peak height of aspartic acid in the same sample)
of α-aminobutyric acid, urea, β-alanine, citrulline, α, γ-diaminobutyric, and cysteic acids and ornithine were found. In *Mariscus*
leaf litter the distribution was similar except that ornithine and
cysteic acid had risen to 12.7 and 10.5%, respectively. In the hydrolyzate of the whole peat at the surface level, γ-aminobutyric acid,
not found in fresh plant or litter, appeared with a relative peak
height of 12%, and β-alanine had increased. The data in Table 6 show
the rich and varied abundance of nonprotein acids found in some pH-1
soluble fractions and fulvic acid hydrolyzates. The studies by Mathur
and Paul [188] with *Penicillium* indicate that nitrogen bound to humic
acids is at least partially available to microorganisms, but it seems
probable that once amino acids are complexed with insoluble materials
they can only be released slowly to satisfy nutritional needs. Because the concentration of free acids in these and other peats is so
low, the rate of complexing must be quite high. Therefore, it is the
free acids -- and apparently to some extent those bound to fulvic
acids -- that reflect recent metabolic activity. The appreciable
amounts of α-aminobutyric acid, urea, β-alanine, citrulline, and
ornithine no doubt reflect this activity.

γ-Aminobutyric acid can be formed as a microbial breakdown product of glutamic acid. However, its very high concentration among
the free acids in the surface level of the *Mariscus* site together
with the appreciable amounts of α, γ-diaminobutyric and α, ε-diaminopimelic acids, no doubt derive from bacterial cell walls, as may some

of the ornithine. The "miscellaneous" acids (sarcosine, norvaline, norleucine, and the "unknown peak") sometimes represent substantial contributions, but more often do not.

TABLE 6

Nonprotein Amino Acids and Urea in Some pH-1
Soluble Fractions and Fulvic Acid Hydrolyzates[a]

| Amino acid | Mariscus site[b] | | | | Mangrove site[b] | |
| | pH-1 sol. | | fulvic acid | | pH-1 sol. | fulvic acid |
	0 ft	6 ft	0 ft	6 ft	7 ft	7 ft
α-Aminobutyric	19	33	n.f.	6	35	47
Sarcosine	n.f.	n.f.	tr.	2	6	n.f.
Urea	19	22	1	1	7	11
Norvaline	n.f.	n.f.	n.f.	2	7	n.f.
β-Alanine	13	34	39	23	27	21
Norleucine	n.f.	30	n.f.	2	10	7
γ-Aminobutyric	260	6	10	13	67	6
Citrulline	32	2	2	1	8	13
Glutamine	1	n.f.	tr.	n.f.	19	n.f.
α, γ-Diamino-butyric	117	4	5	13	6	3
Ornithine	68	4	3	17	38	9
Cysteic	116	7	3	8	15	19
α, ε-Diamino-pimelic	7	10	5	6	75	8
Unknown peak[c]	28	tr.	6	tr.	n.f.	6
Aspartic acid[d]	1.8	0.3	340	106	0.1	7.5

[a]Relative peak areas: aspartic acid = 100; for absolute concentration of aspartic acid, see last line of table.
[b]n.f. = not found; tr. = trace found.
[c]Unknown peak: Had a reproducible retention time; experiments with known compounds showed that canavanine, homoserine, and α-aminocaprylic acid all shared this retention time.
[d]Absolute concentration in μg/g dry ash free peat.

The increase in amino acids with depth noted by Swain [16] and by Krzeczkowska and Szczepaniak [215] is not found in the Everglades peats or any of their fractions; in these almost all concentrations and the nonprotein/protein acid ratios varied in a random manner.

The conclusion reached was that the detailed analyses at any one level reflected the complex set of processes that prevailed when that level constituted the surface zone and little change occurred after burial. There was some evidence that amino acids in the humin fraction tended to become more tightly bound with increasing depth and to require stronger acid and more time for hydrolysis [207].

As in other peats [164, 205, 213], Casagrande [207] found amino sugars in the Florida samples, although in quite small amount (0.2-0.7 mg/m); in the *Mariscus* peat there was some tendency for them to be concentrated in the fulvic acid hydrolyzate. These substances constitute the monomer unit of chitin, the structural material of fungal cell walls and of the exoskeletons of insects and molluscs. Glucosamine is also an important constituent of the polymeric muco-complex of bacterial cell walls.

E. Carbohydrates

Essentially all articles that give compositional data on peats include proximate analyses containing entries for "cellulose" and "hemicellulose," or "difficultly hydrolyzed polysaccharides" and "readily hydrolyzed polysaccharides." Some of these result from applying to peat the methods used for fractionating fresh plant material, and some of them depend on fractionation schemes specially devised for peat; methods of fractionating peat have been reviewed [156, 157, 164]. We are not satisfied that all of the materials reporting in the classes named are indeed plant polysaccharides. The reported contents of difficultly hydrolyzed material range from 2-30% in various peats, and of readily hydrolyzable material from 5-50% of total organic matter. Where it has been measured, there is a moderately good correlation between the degree of decomposition and the polysaccharide contents, the latter decreasing as the former increases. There are, accordingly, correlations with peat type, high moor peats tending to have more polysaccharides than low-moor and forest types.

Rakovskii et al. [25] give some figures that we find illuminating.

They separated a *Sphagnum* peat from Russia and a forest peat from
Ceylon by sieve analysis into "fiber" and "humus" fractions, and
then made proximate analyses of both pairs of fractions. Their results are, in part, shown in Table 7.

TABLE 7

Proximate Analyses of Size Fractions of Two Peats[a]

Percent in peat	Fiber		Humus	
	USSR Sphag.	Ceylon forest	USSR Sphag.	Ceylon forest
Readily hydrolyzable	23	5.1	35	10.5
Cellulose	14	3.8	2	0.8
Lignin	22	26	9.5	7.7
Humic acids	15	43.4	30	62.4
Fulvic acids	12	15.3	8	12.8

[a]From Rakovskii et al. [25].

The concentrations of cellulose and lignin are considerably
higher in the fiber fraction, which contains the recognizable plant
tissue, than in the humus, as is reasonable. The amounts of these
substances in the humus fraction may well represent single cells and
severely macerated tissue. The greater part of the humic acids is
in the humus, although the order is reversed for fulvic acids; there
is an implication here that humic and fulvic acids can be generated
in cell walls while these are still intact. However, the difficult
question is, why is the greater part of the "readily hydrolyzable
polysaccharide" associated with the humus rather than the fiber
fraction? Data in the same paper indicate that hemicelluloses represent 40-50% of the total readily hydrolyzable polysaccharides;
this is more than the proportion of readily hydrolyzed polysaccharides shown on Table 7 as present in the fiber fraction, implying that
some of the "hemicellulose" must be in the humus fraction and there-
for probably not plant structural polymer at all.

Lucas [157, 216] separated holocellulose, hemicellulose A and
B fractions, and α-cellulose from fresh *Mariscus* plants, organs of
R. mangle, and surface litter from these. Hydrolyzates of the

polymers from fresh plant material, analyzed as alditol acetates by
gas chromatography, showed the expected distributions of sugars.
The cellulose contents of surface litter samples were not much different from those of the corresponding fresh materials, but the
degree of polymerization, determined viscometrically, had dropped
by 20-70%, depending on the sample.

The sugar distributions in hydrolyzates of the hemicellulose A
fractions of the various litter samples were on the whole reasonable
for glucuronoxylan polymers, but in a number of cases the distributions in the hemicellulose B hydrolyzates were not. Thus the sugars
from this fraction of partly decayed *Mariscus* leaves contained
(fresh plant values in parentheses): xylose, 9% of total sugars (57%);
arabinose, 15% (7%); rhamnose, 1.7% (0.2%); fucose, 2.0%, (0.8%);
and uronic acids, 62.4% (8.1%). Similar results were obtained for
the hemicellulose B fraction of the buried stump of *R. mangle* [160],
except that here the increase in fucose was even more striking (0.9%
to 4.1%).

Extraction with 0.5 N sodium hydroxide of partly decayed *Mariscus*
leaf blades and *R. mangle* stems from litter, and of the buried mangrove stump, yielded materials apparently very similar to humic acids.
Acid hydrolysis of these materials afforded mixtures of sugars the
principal components of which were xylose and uronic acids. It
appeared that glucuronoxylans from the plant hemicelluloses had been
so degraded in molecular weight that they become soluble in 0.5 N
NaOH without prior delignification. There was more uronic acid than
would be equivalent to xylose in glucuronoxylan, so that presumably
some polyuronides were also extracted with the humic acids.

The specific rotations, $[\alpha]_D^{25}$, of the structural polysaccharides
of higher plants are all negative, consistent with their possession
of β-glycosidic linkages. The hemicellulose A and B fractions of
partly decayed plant parts were all appreciably less levorotatory
than the polymers from fresh tissue, e.g., for the polymers from
fresh *R. Mangle* stem, $[\alpha]_D^{25} = -78°\alpha$ for A and $-87°\alpha$ for B; from the
buried stump, $[x]_D^{25} = -29°\alpha$ for A and $-7°\alpha$ for B. These changes can

only be accounted for by postulating the presence in the litter fractions of dextrorotatory polymers of microbial origin.

Investigations of polysaccharides in mineral soils indicate that not only decayed plant debris but also the extracellular slimes and capsules of bacteria and fungi and the carcasses of the soil microflora and microfauna, may act as the sources of polysaccharides [219-221]. The microbial carbohydrates are often rich in uronic acids, arabinose, rhamnose, fucose and ribose; these monomers were found in the hydrolysates of the surface litter on the Florida peats. Moreover, a polymer containing 76% uronic acid and small amounts of eight neutral sugars was isolated from the fulvic acid fraction of a fresh water peat; its specific rotation was $+31^{\circ}$C.

In further work Lucas attempted to isolate polymeric α-cellulose from the peats (the two same sites studied by Casagrande; see p. 186). At some levels he could find none; at levels where he did, the concentrations found were 0.3-0.6% of dry organic matter; and the degree of polymerization had dropped to 20% of the fresh plant value in the *Mariscus* peat and to 10% in the mangrove peat. Similar comparisons of various plants and peats in Russia indicated that the degrees of polymerization of α-cellulose in peats are 14-34% of the appropriate values for fresh plant material and that amino acids were firmly bound to α-cellulose [222].

Sugar distributions in hydrolyzates of the Florida peats and their fractions showed that, except at the surface level in the *Mariscus* core, there is little evidence that hemicelluloses survive in peat; xylose has become a minor component, often exceeded in abundance by one or more of galactose, mannose, rhamnose, or ribose [95, 157, 216]. In particular, the fulvic and humic acid hydrolyzates had sugar distributions quite unlike those of hemicelluloses; in those from the mangrove peat, they contained principally mannose as the most abundant sugar. Glucose was the most abundant sugar in the whole peat and in the humin hydrolysates, which may have been derived at least partly from cellulose, but many other sugars were present in significant amounts, including fucose. In the hydrolyzate of the whole peat

from the 5-ft level of the *Mariscus* core, glucose represented 39% of
total neutral sugars, rhamnose 23%. Finally, uronic acids were major
components of the sugars in hydrolyzates, representing 0.5-2% of
total organic matter.

Kadner et al. [205] found arabinose and rhamnose in a number of
peats, and galactose, galacturonic acid and glucuronic acid in some.
Dragunov and Shmelev [223] discussed the relation of uronic acids to
humic acid formation, and Verner [224] believed that polyuronides in
peat were formed by the oxidation of plant polysaccharides, which
could not be the case with the polyuronide isolated by Lucas (above).
He failed to find fucose, rhamnose or ribose in peats, whereas Duff
[225] found the 2-0-methylether of rhamnose in a peat, which he felt
must have had a microbial origin. 2-*o*-methylxylose and 3-*o*-methylxy-
lose were found in a peat derived from *Calluna vulgaris* [226]. The
first of these is known to occur in higher plants, but the second has
not previously been found in nature. The determination of polysaccharides in peat has been discussed by several authors [151, 224,
227], as have fractionation schemes to yield proximate analyses including polysaccharide values [155, 156, 228-231].

Exarhos and Given (unpublished, 1972) have adopted a simple
approach to follow the rate of destruction of cellulose in peats.
They wrapped sheets of filter paper around Teflon or nylon rods some
60 cm long. The paper was protected from abrasion, and the rods were
pushed into peat in various environments. In 5 months, paper a few
centimeters below the surface was completely destroyed. In 3 months
in two red mangrove habitats, paper between about 3 and 20 cm below
the surface was largely destroyed, but that above or in the surface
layer was not. It was possible to isolate microbial colonies from
swollen and weakened paper after 1 or 2 months exposure and these
are being investigated.

F. Lipids and Other Organic-Solvent-Soluble Compounds

The material in peats that is soluble in organic solvents does
not constitute a major fraction of the whole. It is principally of

interest in so far as it can help to identify the sources of the organic matter in the sediment. The fatty acids of lipids have a considerable degree of taxonomic specificity that could be of great value if one had the patience and facilities for distinguishing branched-chain, cyclopropane, hydroxy, and unsaturated acids from straight-chain acids, and if the positions of the double bond(s) were located. No study of any sediment has as yet attempted to distinguish all of these types. Even with the more easily analyzed types there is a useful degree of specificity, particularly among the longer chain acids ($>C_{20}$). The waxes of leaves and fruits contain distributions of hydrocarbons and fatty acids that may be species- or genus-specific [232, 233].

Sterols and triterpenes, and the saturated hydrocarbons derived from them under geologic/geochemical influences, have received considerable attention in recent years from geochemists, because they show a fair degree of taxonomic specificity that is of value in determining the source of organic matter in rocks [234-236].

Bel'kevich et al. [237] in a recent study of a peat wax (probably a *Sphagnum* peat) report that 24.2% of the wax consisted of saturated straight-chain acids, mostly C_{16}-C_{26}, and their esters, whereas 3.6% represented branched chain acids. Although branched chain acids are known in leaf waxes, they are rare, so far as is known. However, the C_{15} and C_{17} iso and/or anteiso acids are common in bacteria, and in some organisms C_{17} and C_{19} branched-chain acids are found [4]. Perhaps the branched-chain acids in the peat wax represent a bacterial contribution. Hydroxy acids were also found; these, presumably ω-hydroxy-alkanoic acids, would have derived from leaf waxes.

A series of straight-chain acids were found in an Irish peat [283]; the acids cover the range C_{18}-C_{34}, with C_{24}, C_{26}, and C_{28} predominating. A strong even/odd carbon number preference was retained. Alkanes appear to have been restricted to C_{29}, C_{31}, and C_{33}. Working without the resources of modern chromatography, Titov [239] claimed to have isolated from *Sphagnum* peat a saturated alcohol, $C_{27}H_{55}OH$, hydrocarbons $C_{33}H_{68}$ and $C_{35}H_{72}$, a hydroxy acid, and esters of cyclic alcohols[*] and cyclic acids. The fatty acids of adipocire, the curious substance formed from animal fats in peat bogs, have been reviewed by Bergmann [240]. Adipocire consists chiefly of palmitic, stearic, oleic, and hydroxystearic acids and their Ca and Mg salts.

The straight-chain acids C_{12}-C_{26} were found, palmitic accounting for 44% of total acids, in the Huleh peat in Israel, a more southerly region; the even-odd predominance was strong [241]. Unsaturated acids (10.8% of total) consisted only of the common acids, $C_{16:1}$, $C_{18:1}$, and $C_{18:2}$. Unless they were generated within the peat, the presence of the unsaturated acids indicates that double bonds can survive the aerobic decay processes on and in the surface.

The sterols, β-sitosterol and β-sitostanol, and the triterpenes, friedelin and friedelan-3β-ol, have been reported in Scottish and Canadian *Sphagnum* peats as well as in the Huleh peat [106, 241-244]. Also found was stigmasterol [243], which is a widely distributed phytosterol in monocotyledons and some dicotyledons (particularly Leguminales and Chenopodiales); it is also found in members of most orders of algae. β-Sitosterol is common in the Bryophyta, which is no doubt why it was found in *Sphagnum* peat. Friedelin is known in members of the Ericales, which may grow in habitats near or with mosses. Admittedly much is still to be known about the chemical taxonomy of these substances, but there are some helpful reviews [234-236].

Sediments examined in lagoons on the Texas coast consisted of a series of superimposed dead algal mats interspersed with layers of mud [245] and may be described as algal peats. The living mat was found to be rich in unsaturated acids, but these progressively decreased in importance with increasing depth. Although some acids not in the living algal mat were found, in general the distribution in the lower levels resembled that in the living mat except for the disappearance of unsaturated acids. It is not clear whether this disappearance was the result of microbial consumption or of aerial oxidation; the latter seems more probable.

---

[*] Sterols? [P.H.G.].

A study of hydrocarbons and fatty acids in living plants, litter
and sediments at four sites in the Florida Everglades was made by
Cooper [4]. The plants were *R. mangle*, *Mariscus jamaicensis*, and
Nymphaea, and the sites, apart from the two studied by Casagrande
and Lucas (see pp. 186 and 193) were a brackish stand of *M. jamaicensis*
and a water lily pond in a freshwater area. Bimodal distributions of
both hydrocarbons and acids were observed in plants and peat at the
four sites. One mode centered on the region C_{16}-C_{19}, and the other
C_{24}-C_{30}. With the hydrocarbons in some fresh plants both modes were
important, but usually the waxy compounds were quantitatively unim-
portant compared with the lower mode. In the peats, however, the
upper mode accounted for a large proportion of the total hydrocarbons
and fatty acids, which tended to become even larger with increasing
depth. Litter presented an intermediate picture. It seems improbable
that microorganisms should selectively remove the compounds of lower
molecular weight. It was therefore suggested that protoplasmic
cells were mostly lysed on decay of the organ, and the metabolic
lipids ($<C_{20}$) lost, whereas the cuticle retained its integrity and
resists attack, thus preserving waxy constituents ($>C_{20}$). Resemblances
between the wax components of peats and precursor plants *(R. mangle,*
M. jamaicensis, Nymphaea) was sporadic. In some cores the hydrocarbon distribution quite resembled that in the plants but the acids
did not, while in the other cases the reverse was true.

It was noted in the analyses of peat compounds that small chromatographic peaks in the range C_{16}-C_{19} appeared that were not due to
saturated straight-chain compounds and were not present in the traces
for fresh plant components. It was thought that these might be iso
or cyclopropane acids or unsaturated acids from bacteria (*cis*-vaccenic
acid, a $C_{18:1}$ acid different from oleic, is found only in bacteria).
There was certainly evidence in the upper range ($>C_{24}$) that a number
of branched chain and unsaturated acids and hydrocarbons from sources
other than the known higher plants had been added to the system, and
the carbon preference index was less in peats than in plants [4].
Microorganisms, particularly under anaerobic conditions, can be

responsible for a number of transformations of fatty acids (saturation of double bonds, β-scission to give acids with two fewer carbon
atoms, complete resynthesis). This has been demonstrated in a
estuarine mud by Eglinton et al. [246] using oleic acid labeled with
both ^{3}H and ^{14}C.

There is no doubt that the study of lipids in peats can be very
informative, but it is also clear that the full facilities of a combined gas chromatograph-mass spectrometer system are needed in a
fully adequate study.

VII. PEAT-FORMING PROCESSES: CONCLUDING REMARKS

In the introduction it was pointed out that peat represents an
escape from the carbon cycle, and that this is, in many ways, its
most important characteristic.

Estimates of productivity in transitional and rheophilous peat
swamps [21, 247-249] indicate that, considering similar plant communities, peat lands are comparable to other types of environments
[250]. Mangrove swamps, in particular, are highly productive [247].
Therefore, as regards the indigenous flora of such undisturbed swamps,
peat is not necessarily a nutritionally impoverished growth medium.
Ombrophilous bogs are probably in a different category in respect of
productivity and may represent more extreme habitats for plant
growth.

In productive swamps we may infer that there are adequate supplies
of nutrients available including nitrogen, phosphorus, and potassium.
As we have seen in earlier sections there are relatively limited
populations of microorganisms present in most peats and this, combined with the limited decomposition of the plant remains, implies
that the economy of these substrates with respect to such nutrients
may differ considerably from mineral soils.

There is some evidence that the rate of decay of peat material
is restricted by a shortage of available nitrogen. This may be
partly caused by an overall deficiency or by a shortage of nitrogen

in forms that may be utilized by microorganisms. Similar apparent
deficiencies of other nutrients may also restrict the growth of saprophytic organisms and hence contribute to the generally low level of
microbial activity. Grosse-Brauckmann and Puffe [249] found that
lowmoor and transitional peats increase annually in depth by amounts
in the range 0.1-1.6 mm, the average being 0.4 mm. If one takes this
mean value and assumes the density of peat to be 1.1, peat is accumulating at a rate of 440 kg/ha/year. If we round this off to 500 kg/
ha/year and further assume that the production of fresh biomass is
5000 kg/ha/year, then it can be calculated that about 10% of the
annual productivity has accumulated as peat. This suggests another
simple but illuminating calculation. Taking Heald's figure [247]
for production of mangrove litter in the North River area of the
Everglades, 0.88 kg/m^2/year, and multiplying it by 3000 years, which
is the approximate age of the basal section, a total of 2.64 ton/m^2
plant debris has been produced during the history of the sediment.
In a 3-m section of the peat, 1 m^2 in area, there is about 0.2 ton/m^2
organic matter. That is, about 7.5% of the production of plant material has accumulated as peat. Thus, the efficiency of the peat-forming
processes is quite low in that most of the plant material produced
is decomposed either during peat formation and/or in deeper horizons.
Such a plant community would require a net nitrogenous input of about
6-12 kg/ha/year which might be supplied by rainfall and the groundwaters,
if it were assured that all N in the peat is available.

We would argue that if chemical inhibition of microbial activity
exists, it is not a universal effect, but selective for particular
tissue-destroying organisms. Some authors have found that peat extracts stimulate the growth of plants [251-255] and influence pigment
production [252, 254]. It has been claimed that the quantity of acetone produced per unit of glucose by *Clostridium acetobutylicum* is
increased if peat extracts are added to the medium [256], but no general study of the effect of peat on microbial growth has been made.
Lewis and Starkey [257] followed the rate of decomposition of plant
polysaccharides in mineral soils and found that it was somewhat reduced if extracts of (somewhat atypical) condensed tannins were added.

Exarhos and Given (unpublished observations, 1971) found no change
in the rate of cellulase-catalyzed production of glucose from filter
paper when macerated mangrove peat was added to the reaction mixture.

It is not clear, therefore, whether the relatively slow and
limited decomposition of plant tissue in peats is caused by chemical
inhibition or by nutritional factors, but the latter is more probable.
It has been claimed that addition of nitrogen enhances the activity
of microorganisms in degrading lignin and cellulose [258] and that
on fertilization of peat and raising of crops on it much previously
unavailable nitrogen is released [259].

The presence of microbially derived compounds, including amino
acids, certain lipid components, amino sugars, certain carbohydrates,
and fixed sulfur, in deep peats provides indirect evidence of the
presence and activity of a microflora. Moreover, dehydrogenase activity has been found at all levels in some peats and only near the
surface in others. If we take into account the hydrolyzable amino
acids, the compounds (presumably also amino acids) represented by
unhydrolyzable nitrogen, some of the carbohydrates and a part of
the humic acid constituents, we see that, quantitatively, microbially
derived substances may make a substantially higher contribution to
the organic matter of a peat than has been realized in the past. This
is in accordance with recent studies of litter in other communities
where the material immediately above the mineral soil is comprised
mainly of microbial structures or remains.

The humification process in peats is generally said to consist
of an initial aerobic phase with disappearance of the surface plant
organs, followed by an anaerobic phase in which the decomposition
rate is lower (e.g., 79). We have encountered only one attempt actually to measure oxygen tensions in peat. Efimov and Efimova [260]
found that the oxygen content of the surface waters of peats varied
from 0.2 to 6 mg/liter, changing with the seasons and from year to
year. Intense photosynthesis in the late summer of favorable years
caused the highest contents, and very low contents were found in
spring. In a transitional peat bog 4 mg/liter was found at the

surface, decreasing to 0.9 mg/liter at the boundary with a gley horizon. Several factors tend to increase oxygen availability: aeration in the rhizosphere by transport down through roots, vertical or horizontal bulk flow of water after heavy rainfall, presence of algae, and partial drainage and replenishment with aerated water in intertidal swamps.

We prefer to consider the humification processes as involving extensive mechanical disintegration of plant organs rather than their disappearance. Apart from the laboratory studies of Kaleja [153] there is little evidence suggesting that anaerobic decomposition of polysaccharides occurs to a significant extent, and it appears that no organism is capable of decomposing lignin anaerobically. Other anaerobic processes of significance do take place, including sulfate reduction and nitrogen fixation.

It is clear that if peat studies are to be fruitful they should in the future be informed by a more exact knowledge of the botany, microbiology, and chemistry of the system. Microscopic examination of tissues should complement biochemical studies, and the latter must take advantage of the full potential of modern instrumental methods of analysis.

REFERENCES

1. A. D. Cohen, Ph.D. Thesis, *The Petrography of Some Peats of Southern Florida (With Special Reference to the Origin of Coal)*, The Pennsylvania State University, University Park, 1968.

2. H. Jacob and J. Koch, in *3rd Intl. Peat. Congr.*, National Research Council of Canada, Ottawa, 39 (1968).

3. C. H. Dickinson and M. J. Dooley, *Plant and Soil*, *27*, 172 (1967).

4. W. J. Cooper, *Geochemistry of Lipid Components in Peat-Forming Environments of the Florida Everglades*, M.S. Thesis, The Pennsylvania State University, University Park, 1971.

5. M. N. Nikonov, *Byoll. MOIP see Geol.* *23*, 93 (1948).

6. A. M. Matveev, M. I. Neistadt and A. S. Olenin, in *3rd Intl. Peat Congr.*, National Research Council of Canada, Ottawa, 382 (1968).

7. C. Q. Weber, *Englers Bot. Jahrb.*, *90* (Leipzig) (1908).

8. D. A. Ratcliffe in *The Vegetation of Scotland*, (J. H. Burnett, Ed.), Oliver and Boyd, Edinburgh, 1964.

9. G. K. Fraser, Paper to Section B2, *Intl. Peat Symposium, Dublin*, 1954.

10. H. Sjors, *Oikos*, *2*, 241 (1950).

11. E. Gorham, *Oikos*, *2*, 217 (1950).

12. S. Mattson, G. Sandberg, and R. E. Terning, *Ann. Agri. Coll. Sweden*, *12*, 101 (1944).

13. M. P. Volarovich, N. T. Korol, I. L. Lishtvan, A. M. Mamtsis, and N. V. Churaev, *Prir. Bolot. Metody Ikh Issled., Mater., Vses. Soveshch. Sovrem. Puti Metody Issled., Bolot.* 1964 (Pub. 1967), 149; *Chem. Abstr.*, *69*, 108503 (1968).

14. S. Kulczynski, *Peat Bogs of Polesia*, *Mem. Acad. Sci., Cracovie* Ser B., 356 (1949).

15. D. J. Bellamy, *Third Internat. Peat Congr.*, National Research Council of Canada, Ottawa, 74 (1968).

16. F. M. Swain, *Non-Marine Organic Geochemistry*, Cambridge Univ. Press, Cambridge, 1970.

17. C. U. Loveless, *Ecology*, *40*, 1 (1959).

18. L. A. Jones, Ed., *Univ. of Florida, Agri. Exp. Sta. Bull.*, *442*, 1948.

19. J. H. Davis, *Florida Geol. Survey Bull.*, *30*, 1 (1946).

20. W. Spackman, D. W. Scholl and W. H. Taft, *Field Guide Book to the Environments of Coal Formation in Southern Florida*, Geol. Soc. Amer. 1964.

21. W. G. Smith, Ph.D. Thesis, *Sedimentary Environments and Environmental Changes in the Peat-Forming Area of South Florida*, The Pennsylvania State University, University Park, 1968.

22. R. S. Farnham, in *2nd Intl. Peat Congr.* (Leningrad, 1963), R. A. Robertson, Ed.), Vol. I, p. 115, Her Majesty's Stationery Office, Edinburgh, (1968).

23. P. D. Varlygin, *Torf Delo.*, *9*, 6 (1924).

24. I. M. Kurbatov, *Second Internat. Peat Congr.* (Leningrad 1963), (R. A. Robertson, Ed.), Vol. 1, p. 133, Her Majesty's Stationery Office, Edinburgh (1968).
25. V. E. Rakovskii, V. Baturo and L. Pigulevskaya, *Second Internat. Peat Congr.* (Leningrad 1963), (R. A. Robertson, Ed.), Vol. II, p. 965, Her Majesty's Stationery Office, Edinburgh (1968).
26. C.E. Marshall, *The Physical Chemistry and Mineralogy of Soils* Vol. I, Soil Materials, J. Wiley, New York, 1964.
27. A. Szalay and M. Szilágyi, *Adv. in Org. Geochem. 1968* (P. A. Schenk and I. Havenaar, Ed.), p. 567, Pergamon, Oxford, 1969.
28. A. D. McLaren, *Can. J. Soil Sci., 50,* 97 (1970).
29. P. H. Given, J. E. Imbalzano, D. J. Casagrande, A. J. Lucas, A. Chen, and W. J. Cooper, submitted to *Geochim. et Cosmochim. Acta* (1973).
30. S. M. Manskaya and T. V. Drozdova, *Geochemistry of Organic Substances,* (transl. L. Shapiro and I. Breger), Pergamon, Oxford, 1968.
31. E. I. Tarakanova, *Litol. Polez. Iskop., 2,* 136 (1968), *Chem. Abstr., 69,* 108743 (1968).
32. V. G. Klinger, I. F. Largin and M. V. Popov, *Prir. Bolot. Metody Ikh Issled., Mater., Vses. Soveshch., Sovrem. Puti Metody Issled. Bolot,* 1964 (Pub. 1967), 254; *Chem. Abstr., 70,* 49260 (1969).
33. F. Ya. Saprykin and A. N. Sventikhovskaya., *Mater. Soveshch. Rabot. Labor. Geol. Organiz.,* 19th, Leningrad, No. *7,* 95 (1965); *Chem. Abstr., 66,* 67841 (1967).
34. W. Francis, *Coal,* Arnold, Edinburgh, 1961.
35. M. Szmytowna and T. Latour, *Ann. Pharm. (Poznan), 6,* 97 (1967); *Chem. Abstr., 71,* 83392 (1967).
36. M. Szilágyi, *Magy. Kem. Foly., 77,* 172 (1971); *Chem. Abstr., 75,* 83543 (1971).
37. F. M. Swain and N. Prokopovich, *Bull. Geol. Soc. Amer., 65,* 1183 (1954).
38. F. M. Swain, in *Essays on Paleontology and Stratigraphy* R. C. Moore Commem. Vol. (C. Teichert and E. Yochelson, Eds.), p. 445 (1967).
39. W. Armstrong, *J. Soil Sci., 18,* 27 (1967).
40. A. Rave and Y. Avnimelech, in *4th Intl. Peat Congr.,* Vol. 4, p. 263, Otaniemi, Finland (1972).
41. A. Whitfield, *Limnol. Oceanogr., 14,* 547 (1969).
42. W. Stumm, in *Advances in Water Pollution Research,* (O. Jaag, Ed.), Vol. 1, p. 283, Pergamon, New York, 1965.
43. M. P. Volarovich and N. V. Churaev, in *2nd Intl. Peat Congr.,* (R. A. Robertson, Ed.), Vol. II, p. 819, Her Majesty's Stationery Office, Edinburgh, 1968.
44. M. P. Volarovich, N. I. Gamayunov, and I. I. Lishtvan, in *4th Intl. Peat Congr.,* Vol. 4, p. 219, Otaniemi, Finland, 1972.
45. M. P. Volarovich, I. I. Lishtvan, and N. V. Churaev, *Kolloid. Zh. 25,* 22 (1963); *Chem. Abstr., 59,* 1124 (1963).

46. V. I. Lashnev and N. V. Churaev, *Tr. Kalinin Politekh. Inst.*,
     2, 296 (1968); *Chem. Abstr.*, *72*, 34129 (1970).
47. N. T. Korol, *Prir. Bolot. Metody Ikh Issled.*, *Mater.*, *Vses
     Sovesch.*, *Sovrem. Puti Metody Issled. Bolot.*, 1964 (Pub.
     1967), 165; *Chem. Abstr.*, *69*, 108502 (1968).
48. M. P. Volarovich, N. I. Gamayunov, and K. S. Pantelei, *Kolloid.
     Zh.*, *32*, 672 (1970); *Chem. Abstr.*, *75*, 40816 (1971).
49. P. I. Bel'kevich, L. R. Chistova, and L. F. Strogonova, *Khim.
     Tverd. Topl.*, *3*, 47 (1971); *Chem. Abstr.*, *75*, 80660 (1971).
50. P. I. Bel'kevich and L. R. Chistova, in *2nd Intl. Peat Congr.*
     (Leningrad 1963), (R. A. Robertson, Ed.), Vol. II, p. 909,
     Her Majesty's Stationery Office, Edinburgh, 1968.
51. W. Fischer, G. Schlungbaum, and R. Kadner, in *2nd Intl. Peat
     Congr.* (Leningrad 1963), (R. A. Robertson, Ed.), Vol. II,
     p. 985, Her Majesty's Stationery Office, Edinburgh.
52. P. I. Bel'kevich, L. R. Chistova, and L. F. Strogonova, *Vestsi
     Akad. Navuk Belarus. SSR, Ser. Khim Navuk*, *4*, 29 (1966);
     Chem. Abstr., *66*, 108616 (1967).
53. R. C. Salmon, *J. Soil Sci.*, *15*(4), 273 (1964).
54. I. I. Lishtvan, A. M. Mamtsis, and N. V. Churaev, *Prir. Bolot.
     Metody Ikh Issled.*, *Mater.*, *Vses. Soveshch.*, *Sovrem. Puti
     Metody Issled. Bolot*, 1964 (Pub. 1967), 264; *Chem. Abstr.*,
     70, 59492 (1969).
55. F. D. Ovcharenko and S. A. Gordienko, *Agrokem. Talajtan*, *18*,
     25, (1969); *Chem. Abstr.*, *71*, 52042 (1969).
56. R. S. Solodovnikova, L. A. Skripal'shchikova, and V. V. Sere-
     brennikov, *Tr. Tomsk. Gos. Univ.*, *192*, 113 (1968); *Chem.
     Abstr.*, *73*, 59157 (1970).
57. B. Lakatos, J. Meisel, G. Mády, P. Vinkler, and S. Sipos, in
     4th Intl. Peat Congr., Vol. 4, p. 341, Otaniemi, Finland,
     1972.
58. P. I. Bel'kevich, L. R. Chistova and L. F. Strogonova, *Vestsi
     Akad. Navuk Belarus. SSR, Ser. Khim. Navuk*, *4*, 27 (1971);
     Chem. Abstr., 36124 (1972).
59. F. A. Kovalev and V. R. Bensman, *Dokl. Akad. Nauk Beloruss. SSR,
     11*, 624 (1967); *Chem. Abstr.*, *67*, 118998 (1967).
60. T. Mornsjo, *Bot. Notis.*, *121*, 343 (1968); *Chem. Abstr.*, *70*,
     70265 (1969).
61. S. N. Tyuremnov and I. F. Largin, *Tr. Gos. Gidrol. Inst.*, No.
     135, 223 (1966); *Chem. Abstr.*, *68*, 53143 (1968).
62. V. A. Kovalev, *Dokl. Akad. Nauk Beloruss. SSR*, *10*, 477 (1966);
     Chem. Abstr., *66*, 12860 (1967).
63. V. A. Kovalev and V. A. Generalova, *Vestsi Akad. Navuk Belarus.
     SSR, Ser. Khim. Navuk*, *4*, 104 (1968); *Chem. Abstr.*, *70*,
     49259 (1969).
64. Sh. Zhorobekova and A. B. Bugubaev, *Mater. Nauch. Korf.*, *Posvy-
     ashch. 100- [Sto] Letiyu Period.* Zakona D. I. Mendeleeva
     1969 (Pub. 1970), 61; *Chem. Abstr.*, *76*, 5578 (1972).
65. P. H. Given and R. N. Miller, in *Symposium on Sulfur in Coals*
     (A. Cohen, Ed.), Geol. Soc. Amer. Special Pub. (accepted
     for publication, 1972).

66. M. V. Ivanov, *Microbiological Processes in the Formation of Sulfur Deposits*, transl. publ. U.S. Dept. of Agriculture and Nat. Sci. Foundn., 1964 (transl. publ. 1968), 16.

67. A. V. Kochenov and V. N. Kreshtapova, *Geokhimiya*, *3*, 330 (1967); *Chem. Abstr.*, *67*, 4902 (1967).

68. S. L. Shvartsev, *Geol. i Geofiz.*, *Akad. Nauk SSSR, Sibirsk. Otd.*, *7*, 3 (1965); *Chem. Abstr.*, *63*, 14552 (1965).

69. M. G. Savchuk, *Mikroelem. Biosfere Ikh Primen. Sel'skokhoz. Med.*, *Sib. Dal'nevost.*, (1967), 191; *Chem. Abstr.*, *69*, 108504 (1968).

70. S. E. Priemskaya, *Pochvovedenie*, *4*, 59 (1969); *Chem. Abstr.*, *71*, 23681 (1969).

71. R. I. Davies, M. V. Cheshire, and I. J. Graham-Bryce, *J. Soil Sci.*, *20*, 65 (1969).

72. I. F. Largin, S. E. Priemskaya, A. N. Sventikhovskaya, and S. N. Tyuremov, in *4th Intl. Peat Congr.*, Vol. 4, p. 77, Otaniemi, Finland, 1972.

73. M. Deul, Am. Chem. Soc., *Div. Gas Fuel Chem.*, *Preprints*, p. 169 (Sept. 1958).

74. B. R. Gladilovich, G. G. Antonova, N. P. Vard'ya, R. I. Kurbatova, E. V. Mikhailyk, and L. I. Khristoforova, *Zap. Leningrad. Sel'skokhoz. Inst.*, *160*, 4 (1971); *Chem. Abstr.*, *76*, 58142 (1972).

75. E. Valdek, *Tead. Toode Kogumik, Eesti Maaviljeluse Maaparanduse Tead. Uurimise Inst.*, *12*, 131 (1968); *Chem. Abstr.*, *72*, 54164 (1970).

76. A. Toth, *Second Internat. Peat Congr.* (Leningrad 1963) (R. A. Robertson, Ed.), Vol. II, p. 721, Her Majesty's Office, Edinburgh, (1968).

77. S. T. Voznyuk, T. Y. Korobchenko, and N. N. Skochinskaya, *Sov. Soil Science*, (1), 12 (1964).

78. E. Küster and J. J. Gardiner, in *3rd Intl. Peat Congr.*, p. 314, National Research Council of Canada, Ottawa, 1968.

79. H. vanDijk, J. van der Boon and P. Bockel, in *3rd Intl. Peat Congr.*, p. 334, National Research Council of Canada, Ottawa, (1968).

80. F. Maciak, A. Maksimow and S. Liwski, *Second Internat. Peat Congr.* (Leningrad 1963), (R. A. Robertson, Ed.), Vol. II, p. 919, Her Majesty's Stationery Office, Edinburgh, (1968).

81. M. Fiuczek, *Rocz. Nauk Roln.*, Ser. F, *78*, 195 (1971); *Chem. Abstr.*, *75*, 139722 (1971).

82. K. K. Lebedev, *Prir. Bolot. Metody Ikh Issled.*, *Mater.*, *Vses. Soveshch.*, *Sovrem. Puti Metody Issled. Bolot*, 1964 (Pub. 1967), 130; *Chem. Abstr.*, *70*, 86698 (1967).

83. G. Grosse-Brauckmann and D. Puffe, in *2nd Proc. Intl. Working-Meeting Soil Micromorphol.*, Arnhem, Neth., 83 (1964).

84. W. Riegel, *Palynology of Environments of Peat Formation in South-Western Florida*, Ph.D. Thesis, The Pennsylvania State University, University Park, 1965.

85. W. Spackman, C. P. Dolsen, and W. Riegel, *Palaeontographica*, Abt. B, *117*, 135 (1966).

86. W. Spackman, W. Riegel, and C. P. Dolsen, in *Environments of Coal Deposition*, (E. C. Dapples and M. E. Hopkins, Eds.), Geol. Soc. Amer. Spec. Paper, 114 (1969).

87. W. Spackman and E. S. Barghoorn, in *Coal Science*, Amer. Chem. Soc. Advances in Chemistry Series, No. *55*, 695 (1966).

88. R. Thiessen, *Structures in Paleozoic Bituminous Coals*, U.S. Bur. Mines Bull. No. 177 (1920).

89. R. Thiessen, *What is Coal?*, U.S. Bur. Mines Inf. Circ. 7397 (1947).

90. E. S. Barghoorn, *Papers of Robert S. Peabody Foundation for Archeology*, *4*, 49 (1949).

91. E. S. Barghoorn, *Bot. Mus. Leaflets Harvard Univ.*, *14*, No. 1, 1 (1949).

92. E. S. Barghoorn, *Sed. Petrol.*, *22*, 34 (1952).

93. H. Jacob, *Geol. Rundschau*, *51*, 530 (1961).

94. S. A. Musyal, *Sov. Geol.*, *8*, 134 (1965).

95. P. H. Given, in *Advances in Organic Geochemistry 1971* (H. Wehner and H. R. von Gaertner, Eds.), p. 69, Pergamon, Oxford, 1972.

96. J. M. Stewart and E. A. C. Follett, *Can. J. Bot.*, *44*, 421 (1966).

97. V. P. Tropin, *Novye Fiz. Metody Issled. Torfa*, Sb., 1960, p. 37.

98. P. J. Dart, R. J. Roughley and M. R. Chandler, *J. Appl. Bact.*, *32*, 352 (1969).

99. M. P. Volarovich and V. P. Tropin, *Mikrobiologiya*, *32*, 281 (1963); *Chem. Abstr.*, *62*, 6308 (1965).

100. W. C. Elsik, *Micropaleontology*, *12*, 515 (1966).

101. S. Goldstein, *Ecology*, *41*, 543 (1960).

102. C. G. C. Chesters, in *The Ecology of Soil Fungi*, (D. Parkinson and J. S. Waid, Eds.), Liverpool Univ. Press, 1960.

103. P. M. Latter and J. B. Cragg, *J. Ecol. 55*, 465 (1967).

104. R. S. Clymo, *J. Ecol. 53*, 747 (1965).

105. E. Kox, *Archiv. Mikrobiol.*, *20*, 111 (1954).

106. W. A. P. Black, W. J. Cornhill, and F. N. Woodward, *J. Appl. Chem.*, *5*, 484 (1955).

107. V. I. Chastukhin, *Mikologiya i. Fitopatologiya*, *1*, 294 (1967).

108. K. Paarlahti and U. Vartiovaara, *Metsätiet Tutkimuslait. Julk*, *50*, 1 (1958).

109. A. J. Holding, D. A. Franklin, and R. Watling, *J. Soil Sci.*, *16*, 44 (1965).

110. J. H. Baker, *Brit. Antartic Surv. Bull.*, *23*, 51 (1970).

111. P. M. Latter, J. B. Cragg, and O. W. Heal, *J. Ecol.*, *55*, 445 (1967).

112. S. A. Visser, *Life Sci.*, *3*, 1061 (1964).

113. S. A. Waksman and K. R. Stevens, *Soil Sci.*, *28*, 315 (1929).

114. T. G. Zimenko, *Mikroflora Pockv. Severnoi i. Srednei Chasti SSSR*, p. 136, Moscow, 1966.

115. E. N. Mishustin and V. K. Shil'nikova, *Biological Fixation of Atmospheric Nitrogen*, The Pennsylvania University Press, University Park, Pa. and London, 1971.

116. H. de Barjac, VIe *Cong. Int. Sci. Sol*, Rapp. C, Paris, 281 (1956).

117. P. J. Christensen and F. D. Cook, *Can. J. Soil Sci.*, *50*, 171 (1970).

118. Z. Tesic, M. Todorovic and V. Bogdanovic, *Zemljiste Biljka*, *15*, 353 (1966); *Chem. Abstr.*, *68*, 88906 (1968).
119. H. J. Rehm and G. Sommer, *Zbl. Bakt.*, II, Abt. *115*, 594 (1962).
120. S. Ishizawa and M. Araragi, *Soil Sci. and Plant Nutr.*, *16*, 110 (1970).
121. L. I. Skrypka and Z. A. Lysenko, *Mikrobiol. Zh. Akad. Nauk Ukr. RSR.*, *27*, 20 (1965).
122. J. G. Boswell and J. Sheldon, *New Phytol.*, *50*, 172 (1951).
123. J. J. Moore, *Sci. Proc. Roy. Dublin Soc.*, *26*, 379 (1954).
124. C. H. Dickinson and M. Dooley, *Proc. Roy. Irish Acad.*, *68*, 109 (1969).
125. V. I. Chastukhin, *Bot. Zh.*, *52*, 214 (1967).
126. R. W. G. Dennis, *Irish Naturalists' J.*, *13*, 83 (1959).
127. F. Kotlaba and J. Kubicka, *Ceskâ Mykologie*, *14*, 90 (1960).
128. V. I. Chastukhin, *Bot. Zh.*, *51*, 161 (1966).
129. D. Zaborowska, *Acta Mycol.*, *1*, 31 (1967).
130. M. Christensen and W. F. Whittingham, *Mycologia*, *57*, 882 (1965).
131. H. Stenton, *Trans. Brit. Mycol. Soc.*, *36*, 304 (1953).
132. M. I. Timonin, *Canadian J. Res.*, Ser. C, *13*, 32 (1935).
133. N. van Meter, *Some Quantitative and Qualitative Aspects of Periphyton in the Everglades*, M.S. Thesis, University of Miami 1965.
134. P. J. Gleason, *The Origin, Sedimentation and Stratigraphy of a Calcitic Mud Located in the Southern Fresh Water Everglades*, Ph. D. Thesis, The Pennsylvania State University, University Park, 1972.
135. W. Loub, *Sitzungsber. Akad. Wiss. Wien*, Math.-Nat. Kl. Abt. 1, *162*, 545 (1953).
136. K. Leher, *Ber. Bayerischen Bot. Ges.*, *32*, 48 (1958).
137. V. S. Sheshukova-Poretskaya, *Uch. Zap. Leningradsk Gos. Univ.*, *313*, 137 (1962).
138. J. Hayward, *J. Ecol.*, *45*, 947 (1957).
139. G. Marcuzzi and A. M. Lorenzoni, *Stud. Trentini Sci. Nat.*, Sez. B, *46*, 247 (1969).
140. N. N. Smirnov, *Hydrobiologia*, *17*, 175 (1961).
141. L. S. Kozlovskaya, *Pochvovedeniye*, *8*, 35 (1959).
142. J. D. Stout and W. F. Harris, *Bull. N. Z. D.S.I.R.*, *189*, (1968).
143. J. D. Stout, *Soil Biol. Biochem.*, *3*, 1 (1971).
144. P. V. Glob, *Mosefolket: Jernalderens Mennesker bevaret i 2000 Ar.*, The Bos People, Ballantine Books, New York, 1971.
145. S. A. Waksman and E. R. Purvis, *Soil Sci.*, *34*, 95 (1932).
146. C. H. Dickinson, B. Wallace, and P. H. Given, *New Phytologist*, *73*, 107 (1974).
147. L. E. Casida, *Appl. Microbiol.*, *18*, 1065 (1969).
148. L. Janota-Bassalik, *Acta Microbiologica Polonica*, *12*, 41 (1963).
149. C. H. Dickinson and F. Boardman, *Trans. Brit. Mycol. Soc.*, *55*, 293 (1970).
150. P. M. Latter and O. W. Heal, *Soil Biol. Biochem.*, *3*, 365 (1971).
151. L. E. Casida, A. Klein and R. Santoro, *Soil Sci.*, *98*, 371 (1964).
152. E. Yu. Polyakova, *Izv. Vyssh. Ucheb. Zaved., Les. Zh.*, *13*, 146 (1970); *Chem. Abstr.*, *75*, 4647 (1971).

153. E. Kaleja, *Mikrobnyi Sin. Biol. Vazhnykh Veshchestv*, 123 (1968); *Chem. Abstr.*, *71*, 122351 (1969).
154. E. Küster, in *2nd Intl. Peat Congr.* (Leningrad 1963), (R. A. Robertson, Ed.), Vol. II, p. 945, Her Majesty's Stationery Office, Edinburgh, 1968.
155. R. Thiessen and R. C. Johnson, *Ind. Eng. Chem.*, *Anal. Edn.*, *1*, 216 (1929).
156. M. Passer, G. T. Bratt, J. A. Elberling, E. L. Piret, L. Hartman and A. J. Madder, in *2nd Intl. Peat Congr.* (Leningrad, 1963), (R. A. Robertson, Ed.), Vol. II, p. 841, Her Majesty's Stationery Office, Edinburgh, 1968.
157. A. J. Lucas, *Geochemistry of Carbohydrates in Some Organic Sediments of the Florida Everglades*, Ph.D. Thesis, The Pennsylvania State University, University Park, 1970.
158. U. Gupta and F. J. Sowden, *Soil Sci.*, *97*, 328 (1964).
159. J. F. Seaman, W. E. Moore, R. L. Mitchell, and M. A. Miller, *TAPPI*, *37*, 336 (1954).
160. D. J. Casagrande and P. H. Given, *Geochim. et Cosmochim. Acta*, *38*, 419 (1974).
161. G. Kh. Karasaeva and A. B. Bugabaev, *Mater. Nauch. Konf.*, *Posvyashch. 100- [Sto] Letiyu Period*. Zakona D. I. Mendeleeva 1969 (Pub. 1970), p. 27; *Chem. Abstr.*, (1972).
162. J. F. Imbalzano, *Some Chemical and Biochemical Studies of Peat in Southern Florida*, Ph.D. Thesis, The Pennsylvania State University, University Park, 1970.
163. B. R. Hewitt, *Proc. Linn. Soc. New South Wales*, *92*(415), 266 (1968).
164. W. Naucke, *Chemiker-Zeitung*, *92*, 1 (1968).
165. U. Schaefer, Dissertation, Technische Hochschule of Hannover, Germany 1952.
166. E. Leibnitz and H. Hrapla, *Chem. Technik*, *8*, 135 (1956).
167. W. Flaig, in *Advances in Organic Geochemistry 1971*, (H. R. vonGaertner and H. Wehner, Eds.), p. 29, Pergamon, Oxford, 1972.
168. F. J. Stevenson and J. H. A. Butler, in *Organic Geochemistry: Methods and Results*, (G. Eglinton and M. T. J. Murphy, Eds.), 534, Springer, Heidelberg, 1969.
169. J. P. Martin and K. Haider, *Soil Sci.*, *107*, 260 (1969).
170. A. B. Bugabaev and Sh. I. Isaeva, *Mater. Nauch. Konf.*, *Posvyashch. 100- [Sto] Letiyu Period*. Zakona D. I. Mendeleeva, 1969 (Pub. 1970), 87; *Chem. Abstr.*, *76*, 2819 (1972).
171. S. Sipos, E. Sipos, I. Dékány, F. Szántó and B. Lakatos, *Fourth Internat. Peat Congr.*, Otaniemi, Finland, 4 255 (1972
172. R. I. Morrison, *J. Soil Sci.*, *9*,
173. C. Steelink, *J. Chem. Educ.*, *40*, 379 (1963).
174. E. S. Lukoshko and L. V. Pigulevskaya, *Khim. Tverd. Topl.*, *3*, 39 (1971); *Chem. Abstr.*, *75*, 89857 (1971).
175. N. A. Burges, H. M. Hurst and B. Walkden, *Geochim. Cosmochim. Acta.*, *28*, 1547 (1964).
176. W. Wildenhain and G. Henseke, *Monatsh. Chem.*, *100*, 479 (1969).

177. R. I. Tsareva and T. A. Semenova, *Plastidnyi App. Zhiznedeyatel. Rast.*, 95 (1971); *Chem. Abstr.*, *75*, 139728 (1971).
178. S. M. Manskaya and L. A. Kodina, *Geokhimiya*, 370 (1963); *Chem. Abstr.*, *59*, 3659 (1963).
179. G. Bendz, O. Martensson and L. Terenius, *Acta Chim. Scand.*, *16*, 1183 (1962).
180. H. Morita, *Third Internat. Peat Congr.*, National Research Council of Canada, Ottawa, 28 (1968).
181. W. Flaig, in *Coal Science, Advan. Chemistry Series No. 55*, Amer. Chem. Soc., (1966), p. 58.
182. W. Wildenhain and G. Henseke, *Z. Chem.*, *5*, 457 (1965).
183. W. Naucke, H. V. Laaser and F. N. Tarkmann, *Fourth Internat. Peat Congr.*, Otaniemi, Finland, *4*, 45 (1972).
184. V. E. Rakovskii, V. S. Poznyak, M. A. Rakovskaya and V. S. Shimansky, *Trudy Inst. Torfa, Akad. Nauk B.S.S.R.*, *3*, 79 (1954).
185. S. S. Dragunov, *Khim. Tverd. Topl. 1*, 17 (1967); *Chem. Abstr.*, *67*, 21065 (1967).
186. F. Czapek, *Biochemie der Pflanzen*, Fischer, Jena, German, 3rd. Ed. (1922-5).
187. W. Flaig, in *Coal and Coal-Bearing Strata*, (D. J. Murchison and T. S. Westoll, Eds.), p. 197, Oliver and Boyd, Edinburgh, 1968.
188. S. P. Mathur and E. A. Paul, *Nature (London)*, *212*, 646 (1966).
189. S. M. Hopkinson, *Quart. Rev. Chem. Soc.*, *23*, 98 (1969).
190. A. E. Hillis, Ed., *Wood Extractives*, pp. 133, 159, 191, 317, Academic, London, 1962.
191. A. C. Neish, in *Plant Biochemistry* (J. Bonner and J. E. Varner, Eds.), p. 581, Academic Press, New York, 1965.
192. J. Harborne, Ed., *Biochemistry of Phenolic Compounds*, Academic, London, 1964.
193. A. Chen, *Flavonoid Pigments in the Red Mangrove, Rhizophora mangle L, of the Florida Everglades and the Peat Derived from It*, M.S. Thesis, The Pennsylvania State University, University Park, 1971.
194. E. C. Bate-Smith, *J. Exp. Bot.*, *4*, 1 (1952).
195. T. White, in *Chemistry and Technology of Leather*, (F. O'Flaherty, W. T. Roddy, and R. M. Lollar, Eds.), Reinhold, New York, 1958.
196. A. Russell, *J. Amer. Leather Chem. Assoc.*, *37*, 340 (1942).
197. H. L. Hergert, L. E. von Blaricom, J. C. Steinberg, and K. R. Gray, *Forest Prods. J.*, *15*, 485 (1965).
198. H. L. Hergert, in *The Chemistry of Flavonoid Compounds* (T. A. Geissman, Ed.), p. 533, Pergamon, Oxford, 1962.
199. M. Sogo, T. Ishihara, and K. Hata, *J. Japan. Wood Res. Soc.*, *12*, 96 (1966).
200. C. O. Chichester and T. O. M. Nakayama, in *Chemistry and Biochemistry of Plant Pigments* (T. W. Goodwin, Ed.), pp. 452-5, Academic, London, 1965.
201. J. B. Harborne, *Comparative Biochemistry of the Flavonoids*, Academic, London, 1967.

202. K. D. Sears and R. L. Casebier, *Chem. Communs. Chem. Soc.*, 1437 (1968).

203. K. D. Sears and R. L. Casebier, *Phytochem.*, *9*, 1589 (1970).

204. K. D. Sears and R. J. Engen, *Chem. Communs. Chem. Soc.*, 612 (1971).

205. R. Kadner, W. Fischer, and G. Schlingbaum, *Freiberger Forschungsh.*, 159 (1962).

206. V. E. Rakovskii and I. A. Pal'min, *Torf. Prom.*, *42*, 23 (1965); *Chem. Abstr.*, *A63*, 17082 (1965).

207. D. J. Casagrande, *Geochemistry of Amino Acids in Selected Florida Peats*, Ph.D. Thesis, The Pennsylvania State University, University Park, 1970.

208. F. M. Swain, A. Blumentals and R. Miller, *Limnol. Oceanogr.*, *4*, 119 (1959).

209. F. M. Swain, *Bull. Geol. Soc. Amer.*, *72*, 519 (1961).

210. F. M. Swain, G. Venteris and F. Ting, *J. Sedim. Petrol.*, *34*, 25 (1964).

211. V. E. Rakovskii and I. A. Pal'min, *Tr. Kalininsk. Torf. Inst.* No. 13, 87 (1963); *Chem. Abstr.*, *61*, 10508 (1964).

212. I. A. Pal'min and V. E. Rakovskii, *Tr. Kalininsk. Torf. Inst.* No. 13, 96 (1963); *Chem. Abstr.*, *61*, 15293 (1964).

213. Ch. Ivanov, E. Balabanova-Radonova, D. Ruschev, and N. G. Shervashidze, *C. R. Acad. Bulg. Sci.*, *20*, 803 (1967); *Chem. Abstr.*, *68*, 42129 (1968).

214. Sh. Zhorobekova and A. B. Bugubaev, *Sb. Statei Aspir. Kirg. Gos. Univ.*, No. *3*, 118 (1969); *Chem. Abstr.*, *75*, 89856 (1971).

215. I. Krzeczkowska and S. Szczipaniak, *Ann. Univ. Mariae Curie-Sklowdowska*, Sect. D, (1967) (Pub. 1969), *22*, 91; *Chem. Abstr.*, *71*, 72755 (1969).

216. P. H. Given, D. J. Casagrande, J. R. Imbalzano, and A. J. Lucas, *Proc. Internat. Symp. Hydrogeochem. Biogeochem.* (Tokyo, 1970), Internat. Assoc. Geochem. Cosmochem., 1972.

217. J. M. Bremner, in *Soil Biochemistry*, (A. D. McLaren and G. H. Peterson, Eds.), p. 19, Dekker, New York, 1967.

218. F. J. Stevenson, *Soil Soc. Amer. Proc.*, *20*, 201 (1956).

219. S. A. Visser, *Ann. Institut Pasteur*, *115*, 766 (1968).

220. J. Decau, *Ann. Agron.*, *19*, 65 (1968).

221. U. C. Gupta, in *Soil Biochemistry* (A. D. McLaren and G. H. Peterson, Eds.), Vol. , Chapter 4, Dekker, New York, 1967.

222. V. S. Verner, *Khim. Tverd. Topl.*, *2*, 84 (1970); *Chem. Abstr.*, *73*, 36735 (1970).

223. S. S. Dragunov and F. N. Schmelev, *Guminovye Udobr.*, *Teoriya i Prakt. ikh Primeneniya*, *Dnepropetr. Sel'skokhoz. Inst.*, Pt. II, 23 (1962); *Chem. Abstr.*, *61*, 5413 (1964).

224. V. S. Verner, *Vestsi Akad. Navuk Belarus, SSR, Ser. Khim. Navuk*, *3*, 102 (1967); *Chem. Abstr.*, *68*, 23383 (1968).

225. R. B. Duff, *J. Sci. Food Agri.*, *12*, 826 (1961).

226. J. F. Bouhours and M. V. Cheshire, *Soil Biol. Biochem.*, *1*, 185 (1969).

227. Z. Rozmej and A. Kwaikowski, *Zeszyty Probl. Postepow Nauk Rolniczych*, No. 34, 329 (1962); *Chem. Abstr.*, *60*, 11768 (1964).

228. S. A. Waksman and K. R. Stevens, *Soil Sci.*, *26*, 113 (1929).
229. J. W. Parsons and J. Tinsley, *Soil Sci. Soc. Amer. Proc.*, *24*, 198 (1960).
230. O. Theander, *Acta Chim. Scand.*, *8*, 989 (1954).
231. G. Schlungbaum, *Bergbautechnik*, *18*, 203 (1968).
232. G. Eglinton and R. J. Hamilton, in *Chemical Plant Taxonomy* (T. Swain, Ed.), p. 187, Academic, London, 1963.
233. F. B. Shorland, in *Chemical Plant Taxonomy*, T. Swain, Ed.), p. 253, Academic, London, 1963.
234. W. Henderson, V. Wollrab, and G. Eglinton, in *Organic Geochemistry 1968* (P. A. Schenk and I. Havenaar, Eds.), p. 181, Pergamon, Oxford, 1969.
235. I. R. Hills and E. V. Whitehead, in *Advances in Organic Geochemistry 1966* (G. D. Hobson and G. C. Speers, Eds.), p. 89, Pergamon, Oxford, 1970.
236. M. Streil and V. Herout, in *Organic Geochemistry: Methods and Results* (G. Eglinton and M. T. J. Murphy, Eds.), p. 401, Springer-Verlag, Heidelberg, 1969.
237. P. I. Bel'kevich, L. A. Ivanova, and F. L. Kaganovich, *Vesti Akad. Navuk Belarus. SSR, Ser. Khim. Navuk*, *5*, 111 (1971); *Chem. Abstr.*, *76*, 61247 (1972).
238. M. R. Gilliland and A. J. Howard, in *2nd Intl. Peat Congr.* (Leningrad 1963) (R. A. Robertson, Ed.), Vol. II, p. 877, Her Majesty's Stationery Office, Edinburgh, 1968.
239. N. Titov, *Brennst. Chem.*, *13*, 266 (1932).
240. W. Bergmann, in *Organic Geochemistry* (I. A. Breger, Ed.), p. 503, Pergamon, Oxford, 1963.
241. R. Ikan, G. Stahl, and E. D. Bergmann, *Israel J. Chem.*, *6*, 485 (1968).
242. A. J. Ives and A. N. O'Neill, *Can. J. Chem.*, *36*, 434 (1958).
243. J. McLean, G. H. Rettie, and F. S. Spring, *Chem. and Ind.*, 1515 (1958).
244. R. Ikan and J. Kashman, *Israel J. Chem.*, *1*, 502 (1963).
245. P. L. Parker and R. F. Leo, *Science*, *148*, 373 (1965).
246. M. M. Rheed, G. Eglinton, G. H. Draffan, and P. J. England, *Nature (London)*, *232*, 327 (1971).
247. E. J. Heald, *Production of Organic Detritus in a South Florida Estuary*, Sea Grant Tech. Bull. No. 6, University of Miami, 1971.
248. C. L. Porter, *Ecology*, *48*, 937 (1967).
249. G. Grosse-Brauckmann and D. Puffe, *Trans. 8th Intl. Congr. Soil Sci.*, *(Bucharest)* 1964 (Pub. 1967), *5*, 635.
250. E. P. Odum, *Fundamentals of Ecology*, 3rd. Ed., Saunders, Philadelphia, 1971.
251. W. Flaig, *Landbauforsch. Voelkenrode*, *17*, 17 (1967).
252. S. Prát, in *2nd Intl. Peat Congr.* (Leningrad 1963) (R. A. Robertson, Ed.), Vol. II, p. 537, Her Majesty's Stationery Office, Edinburgh, 1968.
253. S. Prát, in *2nd Intl. Peat Congr.* (Leningrad 1963) (R. A. Robertson, Ed.), Vol. II, p. 607, Her Majesty's Stationery Office, Edinburgh, 1968.

212 P. H. GIVEN AND C. H. DICKENSON

254. E. W. Chipman and F. R. Forsyth, *Can. J. Plant Sci.*, *51*, 513 (1971).
255. W. Flaig and H. Sochtig, *Fourth Internat. Peat Congr.*, Otaniemi, Finland, *4*, 19 (1972).
256. E. Kuster, J. Rogers and A. McLoughlin, *Third Internat. Peat Congr.*, National Research Council of Canada, Ottawa, 23 (1968).
257. J. A. Lewis and R. L. Starkey, *Soil Sci.*, *106*, 241 (1968).
258. W. Flaig, *Landbauforsch. Voelkenrode*, 17(1), 1 (1967).
259. D. Haken, *Rostl. Vyroba*, *18*, 9 (1972); *Chem. Abstr.*, *77*, 4225 (1972).
260. V. N. Efimov and Z. S. Efimova, *Zap. Leningrad. Sel'skokhoz. Inst.*, *137*, 34 (1970); *Chem. Abstr.*, *75*, 101139 (1971).

CHAPTER 5

MICROBIOLOGY AND BIOCHEMISTRY OF IRRADIATED SOILS

P. A. Cawse

Environmental and Medical Sciences Division
Atomic Energy Research Establishment
Harwell, Didcot
Berkshire, England

I. INTRODUCTION

Treatment of soil with X- and γ-radiation or an electron beam provides a useful means of partial or complete sterilization for research purposes [1-3], according to the dose applied, that induces very little rise in temperature of the sample. The application of 2.5 Mrad would deposit only 6 cal/g in water and result in sterilization of many soils. In 1956 the International Commission on Radiological Units and Measurements recommended that the imparted energy should be referred to as "radiation absorbed dose," or "rad," corresponding to the deposition of 100 ergs/g of absorbing material [4].

X Rays and γ rays are electromagnetic radiations of identical properties, with wavelengths from 10^{-6} to 10^{-11} cm. The term X rays refers to radiation that arises from the displacement of an internal orbital electron with a beam of high-energy electrons, whereas γ-rays are emitted by nuclei of atoms during decay to a more stable state. The majority of soil irradiation studies have been made with cobalt-60, which emits high-energy γ quanta at 1.17 and 1.33 meV, and this radiation would only be attenuated by approximately 15% after passing through 12.5 cm of water [5]. Caesium-137 has a weaker emission at 0.66 meV and is less suitable for soil work. There is insignificant induction of radioactivity in the irradiated sample since the energy of the γ-photon must generally exceed 1.7 meV before reaction with the atomic nucleus [5, 6]. Sterilization of soil with a 5-meV electron beam can be achieved by exposure of a very thin layer, preferably less than 1.2 cm thick. However, the penetration is poor and 55% of the surface dose is attenuated by 1.6-cm thickness of air-dry soil, having a bulk density of 1.22 g/cm^3 [7].

The availability of radiation sources has greatly improved in recent years with the construction of large installations containing 0.5-1 x 10^6 Ci of cobalt-60 for sterilization of medical goods with

2.5-5 Mrad as a continuous process [8], and packages of soil weighing
up to 25 kg can often be accomodated at a cost of about £2 for 5
Mrad. Several smaller sources have been constructed for the irradiation of food and normally possess a capacity of 0.03-0.25 x 10^6
Ci [9]. On a laboratory scale, 5000 Ci Hotspot sources of cobalt-60
are available [10] and provide dose rates in the region of 1 Mrad/hr
to samples placed in an irradiation chamber of 300 cm^3. Such units
have proved invaluable for studies on the soil response to radiation
and have the advantage that the soil can be withdrawn within seconds
after exposure. A few mobile γ sources have been built for treatment
of potatoes and fruits [11], but would only be sufficiently powerful
to irradiate soil with substerilizing doses not exceeding 1 Mrad. A
patent entitled *Disinfection of Soils* has been filed for a field
irradiation machine [12] and describes the use of either cobalt-60
or an electron generator. The operation of such an apparatus in the
field is doubtful because the use of cobalt-60 in sufficient quantity
to achieve a practicable dose rate would require an extremely heavy
radiation shield to satisfy safety requirements, apart from the poor
penetration of electrons.

Measurement of the radiation dose may be carried out in several
ways. The darkening of color in dyed polymethyl methacrylate (Per-
spex Red 400) on exposure to radiation has been developed as a
dosimetry system accurate to ± 2% over the range 0.5-5 Mrad [13],
with initial calibration of the plastic material by ferrous sulfate
dosimetry [14]. The red plastic dosimeters measure 38 x 11 x 3.2 mm
thick and can be sealed in polythene envelopes for insertion at
different depths in the soil to obtain an accurate dose-distribution
pattern. Silver-activated phosphate glass has also been used for
this purpose in soil columns [15]. In a γ field study by Clark and
Coleman [16] lithium fluoride glass rod microdosimeters were used,
and these became thermoluminescent on exposure [17]. Other workers
studying soil irradiation effects have employed halogenated hydrocarbon dosimeters [18, 19] developed by Sigaloff [20], whereby the
acid liberated from trichloroethylene on irradiation was measured

with chlorophenol red indicator. The change in optical density of
cobalt glass was an effective system to measure doses between 0.001
and 10 Mrad [7, 21].

II. HISTORY OF SOIL IRRADIATION STUDIES

Before the extensive development of powerful X ray machines,
the few pioneering experiments concerned with the response of the
soil microflora to radiation were confined to mixing radioactive
materials either with soil or with pure cultures of soil microorganisms. In 1912, Stocklasa [22] observed that radioactive water
from Joachimstal inhibited *Bacillus mycoides*, *Bacillus fluorescens
liquefasciens*, and *Bacterium* spp., but *Azotobacter chroococcum* was
relatively little affected.

A report of increases in plant yield when pitchblende (uraninite)
was mixed with soil [23] encouraged further work with pure cultures.
Stocklasa [24] exposed *Azotobacter chroococcum*, nitrogen-mineralizing
bacteria, and denitrifying bacteria to α-, β-, and γ radiations from
uraninite and radium. Although it was concluded that β and γ rays
retarded all the organisms examined, α-radiation increased nitrogen
fixation by *Azotobacter*. When air containing the α-particle emitter
radon-222 was passed over the soil, 26% extra fixation was recorded
after 9 months. In contrast, in 1921, Kayser [25] stated that
addition of uranium acetate or phosphate to a culture of *Azotobacter*
was without harmful effect and also improved fixation, and in 1924
it was shown that three species of the bacterium were stimulated to
different degrees in the presence of uraninite [26]. It can be
appreciated that in the early experiments just mentioned dosimetry
was crude or, more frequently, nonexistent. Some 17 years later,
Whelden et al. [27] subjected *Azotobacter* cultures to 0.0005-0.02
Mrad of X rays with only a lethal effect, but studies with a cobalt-
60 source in 1956 by Sokurova [28] led to the conclusion that stim-

ulation of *Azotobacter* only results if irradiation is performed at low intensity, e.g., 0.001 Mrad over 16 hr. A further important observation was that fixation of atmospheric nitrogen is unaffected when cell division is completely prevented by 0.1 Mrad. In other words, the cells have merely been inactivated with respect to proliferation, and are still metabolically active. Recent work on soil respiration [29, 30], nitrate reduction [31], and the bacterial oxidation of ammonium [32] after irradiation has shown a similar behavior, and emphasizes the radioresistance of enzyme systems.

From 1947, with the development of research reactors followed by larger installations for power production, the manufacture of cobalt-60 by neutron activation of the 59 isotope became very economical. It eventually permitted far more extensive investigations on the possible application of high-intensity γ sources in agriculture, medicine, and industry. The development of reactors prompted some limited studies on the effect of radioactive waste disposal to the ground [33, 34], but a significant amount of information on the soil response to radiation did not appear until after 1959. The majority of contributions have been almost equally divided between American and European (including United Kingdom and U.S.S.R.) research efforts. In the main, soil has been exposed to radiation of high intensity for a short period (acute treatment), but some research in γ fields has involved the application of low-intensity radiation over a matter of days or months (chronic treatment). Unless otherwise stated, the radiation treatments mentioned in the following discussion have been applied on an acute basis.

The effect of radiation on soil invertebrates is outside the scope of this review but an idea of their sensitivity relative to the microflora is required. Work by Edwards [35] with a Rothamsted woodland soil showed that many insects were killed by 0.001-0.05 Mrad, and 0.2 Mrad was lethal to all invertebrates by 80 days after exposure. Radiation had a selective action whereby the most active animals showed greater susceptibility.

III. GENERAL SURVIVAL OF MICROORGANISMS IN IRRADIATED SOIL

The processes of sterilization by radiation and heat both follow an exponential law [36]; therefore, the initial density of the soil microflora influences the degree of sterility obtained at a given dose. This effect was originally expressed by Lea [37] as

$$N/N_0 \quad = \quad \exp(-KD)$$

where N is the number of organisms from an initial population N_0 that survive a dose D, and K is a constant that is specific for each microbial species K_a, K_b --- K_i. K will vary according to the influence of the soil environment on the radiosensitivity of the organism [7], and several of these factors will be discussed shortly. In order to calculate the number of organisms that survive after a dose D has been applied to a heterogeneous soil population of total numbers N_t, one would have to know the K constant for each species present, K_a, K_b --- K_i, as well as the total number N_{0t} of the separate organisms initially present N_{0a}, N_{0b} --- N_{0i}. McLaren et al. [7] presented the following equation to describe this relationship

$$N_t/N_{0t} \quad = \quad \overset{i}{\underset{}{\Sigma}}\, N_{0i}/N_{0t} \quad \exp(-K_i D)$$

but the authors pointed out the practical impossibility of using it to predict the dose required to completely sterilize a particular soil. In addition, it can be appreciated that the common practice of air-drying a soil before irradiation could alter the apparent effectiveness of γ treatment owing to a decrease in total numbers of viable microbes [38, 39], as well as from a possible change in the spectrum of organisms present after dessication. This feature is important when comparisons of the survival of microbial populations in soils irradiated in wet or dry conditions are attempted.

Although 2.5 Mrad is normally used to sterilize medical goods
in the United Kingdom [40] and has frequently proved capable of
sterilizing many soils [21, 41, 42], it must be appreciated that
the microbial populations to be eliminated in soil are much larger,
perhaps 10^8 bacteria, 10^5 fungi, and 10^6 actinomycetes per gram.
In experiments for which sterility is essential, for example, as
studies on the relative importance of adsorption and microbial
degradation of herbicides and pesticides in soil, the absence of
viable organisms must be confirmed by standard tests after irradiation. In other cases it may only be necessary to inactivate a
specific organism or groups of organisms such as nitrifying bacteria,
and in such an event either culture tests can be made or the overall
chemoautotrophic activity measured after irradiation. McLaren et
al. [7] did not detect viable organisms in three soils treated with
4 Mrad and incubated with culture media, and in Dublin clay loam
exposed to 1 Mrad the survival of aerobic bacteria and streptomyces
was only 0.001%.

Boyer et al. [43] enumerated the viable microflora before and
after treatment of a wide range of 15 soils from Europe and Africa
with 4 Mrad, and reported 0.0013% survival in the soil possessing
the highest initial population, amounting to 81 x 10^5 total microflora per gram. A lower percentage viability was found in calci-
ferous soils compared with ferruginous ones, where up to 0.06% of
the initial population survived. However, the nature of the microflora could have differed between the soils and would explain part
of this variation in radiosensitivity. However, a suspension of
Serratia marcescens, a proteolytic soil bacterium, was rendered
more resistant to radiation by the presence of calcium in the media
[44]. Cell membranes were considered to be the likely structures
that were protected, probably by stabilization of the monolayers
of fatty acids and proteins.

Jackson et al. [45] examined the effect of soil moisture on
survival: soil was irradiated air-dry, or rewetted to 30% moisture
and incubated for 1 week before γ treatment. Fewer organisms

survived in moistened samples in spite of extensive proliferation before irradiation compared with dry soil (Table 1). It is well known that the radiosensitivity of cells increases in relation to their metabolic activity [28, 46], which would be greater in moistened soil with many cells in a log phase [47]. Another possibility is that organisms in a highly aqueous environment are more severely inhibited by indirect action from the breakdown products of radiolysis of water, rather than by direct damage to the cell structure. Excited and ionized water molecules are formed by the absorption of a high-energy photon, and can react with other water molecules to give free radicals (neutral atoms with an unpaired electron in the outer orbit), such as OH and H, and HO_2 if dissolved oxygen is present [48]. These products are extremely reactive with

TABLE 1

Organisms Surviving γ-Radiation in Clarion Clay [45][a]

Radiation dose (Mrad)	Number of organisms per gram dry soil			
	Bacteria and actinomycetes[b]		Fungi[b]	
	Air-dry	Wet	Air-dry	Wet
0	840,000	7,200,000	8,800	17,800
0.04	311,000	280,000	3,000	1,200
0.1	106,000	118,000	600	160
0.2	17,000	80,000	450	20
0.3	7,000	15,700	200	4
0.6	3,460	5,560	4	0
0.8	1,820	1,132	0	0
1.0	58	122		
1.7	40	0		
2.2	2	0		
2.7	0	0		
3.1	0	0		

[a] pH 5.5, 1.71% organic C, wet soil = 30% water by weight.

[b] Bacteria plus actinomycetes counted on egg albumen agar, and fungi on rose bengal-streptomycin agar.

solute molecules and exist for about 10^{-3} sec [49]. In the presence of pyrimidine bases of nucleic acid, such as thymine and uracil, the oxidizing radicals can form organic hydroxyhydroperoxides [50]. Compounds of this class were also obtained by Okada [51] on treatment of amino acid solutions with 0.05 Mrad. Kashkina and Abaturov [52] irradiated soil at 60% water-holding capacity (WHC) or after air-drying for 4 months, and the percentage of viable fungi, actinomycetes, and bacteria was again higher in dry soil. It is reasonable to expect that a microbial population selected by its resistance to drying and that in the course of which selection has probably produced drought-resistant spores, is also resistant to radiation.

Spores and cysts of *Azotobacter* are more radioresistant than vegetative cells [18, 53], and Monib and Zayed [54] found that doses up to 1 Mrad caused less reduction in the numbers of viable clostridial spores than in total bacterial counts. With regard to the resistance of spores, it is worthwhile to mention pure culture experiments by Vinter [55]: five times more cystine was present in spores of *Bacillus cereus* and *Bacillus megaterium* than in vegetative cells, and it was later shown that the radioresistance of sporulating cells appeared simultaneously with the formation of cystine structures localized in the integumental fraction of spores [56].

There is little information concerning the influence of soil organic matter on the lethality of radiation toward the microflora, although work with organisms in pure cultures indicates that some high molecular weight organic compounds have a protective effect that may be partly explained by their reactivity with free radicals [57]. Jackson et al. [45] reported that actinomycetes and bacteria from an upper horizon containing 1.7% organic C were more radioresistant than organisms in lower horizons with 0.2 and 0.5% organic C, but qualitative differences in populations may have contributed to this effect.

The influence of temperature and degree of aeration during γ-treatment on the survival of soil organisms does not appear to

have received attention. Protective effects of freezing and lack
of oxygen are reported for pathogenic bacteria [58], and the food
spoilage yeast *Saccharomyces cerevisiae* was shown to be more damaged
by irradiation at 45°C than at 20°C [59]. The presence of ozone
in the soil air, although not yet demonstrated, may enhance microbial damage, and the toxicity of ozone at the cellular level has
been reviewed by Menzel [60]. Exposure of air to 1 Mrad can yield
18 ppm by volume of ozone and 1-3 ppm has a germicidal effect on
organisms that attack food [61].

Muller and Schmidt [62] avoided the problem of trying to assess
radiosensitivity in a complex soil environment by using a clay model
system. Two strains of *Azotobacter* gave a different response to
X-radiation but both were more resistant when cultured in montmorillonite than in suspension. It was thought that the clay ad-
sorbs amino acids excreted by *Azotobacter*, protecting these compounds against attack by oxidizing radicals that would lead to deamination [63].

The cation-exchange capacity (CEC) of bentonite and kaolinite
is little affected up to 1000 Mrad [64], and 2 Mrad applied to
ammonium-saturated illite, kaolinite, montmorillonite, and vermic-
ulite did not alter the amounts of KCl-exchangeable ammonium [65].
Eno and Popenoe [41] reported that 8 Mrad had no effect on the CEC
of a fine sand and an organic soil.

IV. RESPONSE OF SPECIFIC MICROBIAL GROUPS AND ENZYMES TO RADIATION

A. Fungi

In general, fungi are the most radiosensitive group of soil
microorganisms [7, 41, 45, 52], a feature well indicated by the
survival data in Table 1. Stotzky and Mortensen [66] also showed
that fungi in Rifle peat were increasingly damaged over the range
0.008-0.25 Mrad, but bacteria were little affected. Roberge and
Knowles [67] compared the lethality of radiation toward fungi and

bacteria in pine humus and clearly demonstrated the greater susceptibility of fungi. *Penicillium* spp. and *Trichoderma* spp. are particularly sensitive [52] and 0.0025 Mrad was sufficient to reduce
viability; in contrast *Cladosporium* spp. are relatively radioresistant [52, 68, 69]. Nevertheless, 0.25 Mrad was sufficient to
kill 98% of viable fungal propagules in a Tennessee arable soil
[69], and the few fungi that survived 0.75-1 Mrad were *Phoma* sp.,
Cladosporium herbarum, *Saprolegnia* sp., *Chloridium apiculatum*, and
Trichocladium sp. When the *Phoma* and *Cladosporium* organisms were
cultured and exposed to chronic irradiation levels from 100 to
1375 rad/hr for 21 days in comparison with two radiosensitive species,
Myrothecium roridum and *Aspergillus terreus*, the difference in γ
response was again similar to that in the soil previously treated
with an acute dose [69].

The plant pathogen *Plasmodiophora brassicae* was killed by exposure of soil to 0.05 Mrad [70], and Mosse et al. [71] found that
0.8 Mrad applied to silty clay loam was sufficient to eliminate the
indigenous *Endogone* spp. population, a vesicular-arbuscular mycor-
rhiza infecting plant roots and causing beneficial increases in
growth. When *Allium cepa* was grown in such soil it remained non-
mycorrhizal even though the medium was not sterile with respect to
other organisms [72]. These experiments emphasize the valuable
selective property of radiation, an important advantage over heat
and chemical treatments. Johnson and Osborne [69] remarked that
irradiation of Etowah silt loam enabled the isolation of fungi not
normally seen on dilution plates and suggested that doses in the
region of 0.25 Mrad could be used to isolate rare species.

The significance of heterotrophic nitrification by fungi such
as *Aspergillus* spp. and *Penicillium* spp. [73] in an acid tea soil
from East Pakistan was examined by applying 0.05-2 Mrad [74]. Nitrate failed to accumulate in the incubated soil above 0.25 Mrad, a
behavior that indicated the elimination of fungi, because nitrifying
bacteria are capable of oxidizing ammonium in soils exposed to
0.25-0.75 Mrad [75].

Prolonged chronic irradiation of soil with 4-6 rad/day tended
to decrease the fungal population [76]. Gochenaur [77] made a detailed study on 277 taxa of fungi in oak-pine forest soil treated
for nearly 7 years with from 2665 rad/20-hr day at 2.5 m from the
source to <1 rad/day at 100 m distance (surface doses); fungi producing highly developed conidiophores were most sensitive and at
the highest exposures only unicellular fungi and forms with rudimentary mycelia were prominent. Radiosensitive types, such as members of the Moniliaceae and Mucoraceous fungi, tended to be replaced
by very melanized and often thick-walled members of the Dematiaceae,
e.g., *Cladosporium* and sterile forms. The yeast population proved
radioresistant and was little affected at 5 m from the source [77],
in agreement with observations by Skou [68].

Witkamp [78] examined the microflora in a Puerto Rican rain
forest exposed to chronic cesium-137 radiation for 92 days. Litter
and soil from the defoliated area within 20 m of the source (0.01-1
Mrad total surface dose) had a similar fungal density to samples
from 60 to 80 m distance (0.0001 Mrad), but the bacterial population
and oxygen uptake was greater under high radiation intensity, presumably from the extra influx of fresh litter. Climatic factors
alone, especially drought, severely checked fungal populations in
nonirradiated soil. Studies at the same location by Holler and
Cowley [79] led to the conclusion that the response of fungi to
chronic radiation should be examined at an organismal rather than
a population level.

Laboratory irradiation of Nevada soil with 0.6 Mrad prior to
atomic bomb tests indicated the resistance of *Stemphylium ilicis*,
Fusarium sp., *Phoma* sp., and *Alternaria tenuis* [80]. Cultures obtained from soil at the Nevada site 2 years after the explosions
(within 1 mile of ground zero) again showed predominance of these
four fungi and ten out of 41 species isolated had very darkly pigmented mycelia or spores, or both [81]. The black spores of
Stemphylium sp. were particularly abundant in culture and also
proved highly resistant to ultraviolet irradiation.

B. Actinomycetes and Bacteria

Organisms in the order Actinomycetales appear to be roughly
intermediate in radiosensitivity between fungi and true bacteria
and were entirely absent after 1 Mrad was applied to soil [57] and
compost [82]. Streptomycetes showed a lower percentage survival
than bacteria immediately after irradiation but proliferated at a
faster rate during incubation and exceeded the numbers present in
nonirradiated soil [54]. Chronic irradiation of actinomycete cultures established in sterile soil proved that *Mycobacterium* sp.,
Nocardia corallina, and *Streptomyces griseus* could survive 0.5 Mrad
delivered over 396 days [68]. Perry [83] isolated actinomycetes
from soil at the Puerto Rican experimental forest that had received
0.003-1 Mrad (surface dose) over 92 days but no damage was evident
from measurement of growth and metabolic activity.

Proliferation of *Azotobacter* and nitrifying bacteria in soil
was virtually eliminated by 0.1 Mrad acute radiation, but when 0.38
Mrad was applied over 7 days (15 hr/day) in a γ field it did not
prove lethal to nitrifiers [84]. It was suggested that the fractionated dose allowed partial recovery of the bacteria. The *Azotobacter* population in an alkaline loam at 50% WHC was lowered by
one-third following 0.1 Mrad [54] but 0.5 Mrad was needed to severely
reduce their numbers in several Texas soils that were irradiated in
a dry state [18].

Mesophilic bacteria in mushroom compost were more radiosensitive
than thermophiles [82] and only gram-positive rods, all resembling
spore-formers, were isolated immediately after 1 Mrad [17, 82].
Blumenfeld [85] and Stanovick et al. [86] have also commented on
the general radiation sensitivity of gram-negative organisms with
particular reference to *Rhizobia,* and Bowen and Cawse [87] found
that peas failed to nodulate when grown in 2 Mrad-treated soil. A
pink gram-positive nonsporing rod of the genus *Brevibacterium* per-
sistently survived 2.5 Mrad, ten viable bacteria per gram of soil
still being detected [88]. Similar pink colonies were noticed by

Rovira and Bowen [89] in a red-brown earth after the same dose. A
highly pleomorphic and previously unknown morphological group of
Bacillus spp. appeared in cultures from irradiated soil [90], and
Peterson [29] remarked that 60-95% of microbes surviving 0.8 Mrad
were represented by *Bacillus* spp.

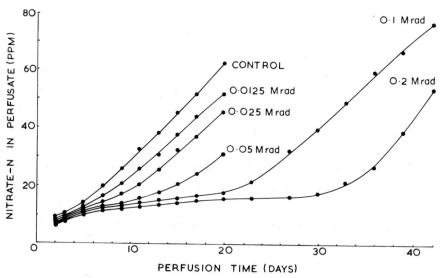

Fig. 1.   Recovery of nitrification in irradiated Grove clay
loam perfused with ammonium sulfate. Reprinted from Ref. [32],
p.397, by courtesy of The Society of Chemical Industry.

The proliferation of nitrifying bacteria toward a state of
bacterial saturation [91] in Grove clay loam perfused with ammonium
sulfate after irradiation is shown in Fig. 1. Cell division was
slightly inhibited by 0.0125 Mrad but was prevented by 0.2 Mrad;
after 25-30 days the survivors began to proliferate at a normal
rate [32]. Furthermore, direct radiation damage to the nitrifiers
appeared to be the major cause of inhibition rather than damage
from toxic compounds produced by radiolysis. Vela [92] described
how proliferation of nitrifying bacteria isolated in liquid culture

from a Texas soil treated with 0.25 Mrad was delayed for 16 days, or for 89 days after 1 Mrad, before rapid oxidation of ammonium commenced.

It must be emphasized that, in spite of the inhibitory effect of radiation on cell division of *Nitrosomonas* and *Nitrobacter*, it is still possible for cell oxidation processes to continue and to result in accumulation of nitrate in many soils exposed to 2.5 Mrad in moist condition [75, 93]. This aspect will be discussed in Section VI, B. Ammonifying bacteria proved slightly less radiosensitive than nitrifiers and were reduced from 22 x 10^6 to 4.5 x 10^6 viable cells per gram of soil by 0.015 Mrad [84]. Recovery was rapid and the numbers were identical to those in an untreated sample by 72 hr after irradiation.

Chronic irradiation in a γ field did not greatly disturb the relative abundance of bacterial groups when 0.074 Mrad was delivered over $3\frac{1}{2}$ months, but the influence of natural fluctuations in temperature and rainfall on the microflora was so great that it masked the radiation effect [94, 95]. Forest-litter soil exposed to 0.8 Mrad over 6 weeks and incubated for an equivalent period showed a 20-fold decrease in bacterial numbers relative to nonirradiated soil, but after acute γ treatment and incubation the population was only reduced by three-fold [16]. This effect is difficult to explain, although Titani et al. [96] found too that *Bacillus subtilis* suspensions were more affected by chronic than acute radiation. However, other studies on inactivation of organisms in culture media have reported the opposite effect or no dose-rate dependence at all [97].

C. Algae

The radiosensitivity of various algae in Arredondo fine sand differed by a considerable extent and whereas some members, notably the green *Chlorophyceae*, were severely damaged above 0.06 Mrad, bluegreen algae tended to be more resistant and even survived 2 Mrad [98]. This finding is supported by pure culture experiments [99]. Following laboratory irradiation of Nevada soil with 1.3 Mrad,

Shields et al. [80] detected *Microcoleus vaginatus*, *Phormidium tenue*, and *Synechoccus cedrorum*; *Anabaena variabilis* and *Nostoc commune* were sensitive and did not survive 0.3 Mrad.

The similarity in radioresistance of certain fungi, actinomycetes, bacteria, and algae in soil is an obstacle to precisely selective elimination of these main groups by increased γ exposure, but research has shown that various doses of radiation can extensively alter the balance both among and within soil microbial groups. It is possible that the predominance of specific organisms or genera can be accentuated by a second γ treatment, following a period of incubation to allow radioresistant types such as *Bacillus* spp. [29, 90] to multiply after an initial dose. Modification of the natural microbial equilibrium by irradiation may be valuable for studies on the role of microorganisms in soil-plant relations, and other possibilities are the selective elimination of plant root pathogens or examination of their pathogenicity in the presence of altered populations of antagonistic nonpathogens. It would be useful to know more about the radiosensitivity of organisms in relation to soil chemical composition.

From observations on the effect of radiation on microbial groups and specific organisms in soils, it is evident that acute or chronic radiation applied to the entire natural microflora can seriously affect the ability of the population to proliferate, respire and transform organic and inorganic compounds and must therefore result in soil biochemical changes of an overall or gross nature.

D. Enzymes

The activity of many soil enzymes shows little decline after a radiation dose sufficient to inactivate all microorganisms [7, 21], and this selective feature has been recognized as of value in research on soil enzyme behavior and kinetics. Phosphatase was only reduced to 37% of its initial activity by treatment of clay loam with 19 Mrad [100] and the activity of urease, saccharase, and

proteolytic enzymes was unaffected by 2 Mrad [42]. The inactivation
of an enzyme by radiation follows the same equation as presented
for microorganisms [7], and McLaren et al. [7] used the decline in
phosphatase activity between 7 and 12 Mrad to calculate a constant
$K = 0.06$. An isolated enzyme, such as potato acid phosphatase, has
a similar K value [100]. Ramirez-Martinez and McLaren [101] reported
that phosphatase was inactivated to a greater extent by raising
the soil moisture content at irradiation. Work in other areas has
indicated that dilution of enzymes increases their inactivation by
radiation and the effect has been discussed by Dertinger and Jung
[57].

Urease in a black spruce humus (60% WHC) was almost completely
inactivated by 4-5 Mrad [102]. However, Skujins and McLaren [103]
stated that urease activity, as measured by a new technique of
$^{14}CO_2$ release from ^{14}C-labeled urea, was increased in some soils
by exposure to 4 Mrad in air-dry condition but decreased in others;
in fact, 8 Mrad was needed to reduce urease activities below their
native level in all soils examined. It was pointed out earlier
[104] that continued enzyme activity in radiation-sterilized soil
could exist in dead cells and could also be situated in or on soil
particles. It is believed that the increase in urease activity
after 4 Mrad results from improved accessibility of substrate to
an intracellular urease component following damage to cell membranes
[103]. In a Texas soil urease activity was stimulated by 2 Mrad
[105]. Soils failing to show extra urease activity after irradiation may possess a predominantly extracellular accumulation of enzyme the activity of which is gradually reduced by increasing radiation dose; in some irradiated soils where this inhibition is more
than compensated for by the extra activity of a mainly intracellular
urease component, stimulation of the enzyme is recorded [103].

Irradiation of urease in black spruce humus at a low dose rate
of 0.24 Mrad/hr resulted in lower activity compared with samples
treated at 1.1 Mrad/hr [102], owing to denaturation of enzyme during
and after irradiation, before assays of activity could be made.
Sterilization of spruce humus by autoclaving completely inhibited

the activity of added urease, whereas 1.1 Mrad was less damaging
[106]; changes in the organic matter during sterilization were
believed responsible. Radiation is much preferred to chemicals for
inactivation of enzymes [107], and a comparative study led Roberge
[108] to conclude that toluene should no longer be used to prevent
microbial proliferation in research on the behavior of soil urease.
The determination of urease was rendered more accurate by exposure
of soil to 1 Mrad because there was no cell division on incubation
of soil-urea-buffer suspension before colorimetric determination of
residual urea [109].

Enzymes involved in respiration, nitrification, and denitrification may all show extra activity in soil that has been irradiated
sufficiently to prevent microbial proliferation and may contribute
to intensive release of CO_2 [30, 110] and disturbance of the nitrogen cycle [32, 75, 93]. The possibility of radiation-induced
breakdown of compounds under heavy radiation treatments normally
used to inactivate enzymes, e.g., 4-12 Mrad or even higher, must
be borne in mind particularly when trying to assess the radiation
response of respiratory and denitrifying enzymes. Some observations
on this subject are discussed in Sections VI,C and VI,D.

V. MICROBIAL RECOLONIZATION OF RADIATION-STERILIZED SOIL

The ability of microorganisms to colonize soil sterilized by
radiation without toxic effects is of importance in the application
of radiation to many aspects of soil research [1, 3]. For investigations on the assimilation of nitrogen by plants, irradiation of
soil is considered superior to heat or chemical sterilization [111].
Barber and Benians [112] developed a technique for culturing
plants under sterile conditions with free drainage and aeration:
the irradiated soil can be inoculated with organisms, such as nitrifying, denitrifying, and nitrogen-fixing bacteria. McLaren and
Skujins [113] reported that the introduction of *Nitrobacter agilis*
into electron-sterilized Yolo silt loam showed no lag or stimulation

in oxidation of nitrite compared with a nonsterile sample. Pure cultures of *Nitrosomonas europaea* and *Nitrobacter winogradskyi* were colonized under conditions of ammonium perfusion in Broad and Black organic calcareous clay loams previously treated with 8 Mrad [114]. When the perfused soil (100 grams) was mixed with 1 kg of γ-sterilized soil, nitrification was well established within 20 days and no nitrite accumulated. Treatment of three soils of pH 7.6-8 with 5 Mrad, followed by inoculation with fresh soil (5 ml of 0.1% w/v suspension per 100 grams irradiated soil), led to the production of 40-70 ppm NO_3-N by nitrifying organisms after 9-weeks' incubation [115].

   Pseudomonas denitrificans readily colonized soil treated with 2-5 Mrad [88], and Peterson [116] proved that *Arthrobacter* sp., *Bacillus megaterium*, *Pseudomonas* sp., *Xanthomonas vesicatoria*, and *Trichoderma viride* grew rapidly in electron-sterilized soil to which glucose had been added after irradiation; however, a delayed log phase was evident when some bacteria were inoculated into auto-claved soil. Further support for this finding came from Salonius et al. [117] working with *Pseudomonas fluorescens* and *Arthrobacter globiformis*, as reproduced in Fig. 2. Their autoclaved soil con-tained more soluble organic matter, soluble carbohydrate, and water-exchangeable electrolyte than the 5 Mrad-treated sample. The tox-icity of two autoclaved soils toward the cotton root rot fungus, *Phymatotrichum omnivorum*, forced Hervey and Williams [118] to use 1.7 Mrad for sterilization of soil in which a culture of the organism was colonized. An enrichment culture of Fe^{2+} oxidizing bacteria was successfully established in soil given 2.5 Mrad for an investiga-tion on the oxidation of pyrite, and the activity of *Thiobacillus ferrooxidans* was evident by the predominance of Fe^{3+} and sulfate in the diffusate compared with mostly Fe^{2+} and less sulfate in non-inoculated soil [119]. Natural reinvasion of γ-sterilized Colum-bia soil in open pots over 21 days gave a 15-fold increase in aero-bic bacteria compared with the numbers in nonirradiated soil [7]. When cores of acid mor soil and associated litter were treated with

5 Mrad and replaced in their natural environment, three species of fungi, *Mortierella isabelliana*, *Pestalotiopsis neglecta*, and *Trichoderma viride*, developed rapidly over 3 weeks [120].

Nevertheless, irradiated soil has not always proved amenable to the reestablishment of organisms. *Myxococcus fulvus* would not

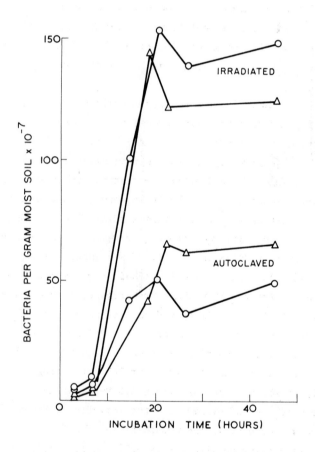

Fig. 2. Comparative growth of *Arthrobacter globiformis* and *Pseudomonas fluorescens* in autoclaved and irradiated soils. Reprinted from Ref. [117], p.245, by courtesy of *Plant and Soil*.

colonize it and grew poorly in an autoclaved sample [116]. Jenkinson et al. [121] observed that inoculation of 2.5 Mrad-treated
or fumigated soil with fresh soil did not reestablish nitrification,
and Merzari and Broeshart [122] failed to colonize nitrifying and
denitrifying organisms in Philippine rice soil given a similar γ
dose. Rovira and Bowen [89] added a soil suspension to red-brown
earth irradiated with 2.5 Mrad and found good colonization of heterotrophic bacteria and fungi, although nitrifying bacteria showed
little activity for 11 weeks. The authors suggested that such
soil, in which the rhizosphere microflora is active, would be useful for studies on the ammonium nutrition of plants.

It is difficult to account for the nature of the inhibitory
factor that is sometimes present in heat- and radiation-sterilized
soils. In 1942, Stanier [123] reported that heat-sterilized glucose solutions inhibited cultures of *Cytophaga* spp., and Meiklejohn
[124] noticed a similar effect on nitrifying bacteria. When hexose
sugars are exposed to 2 Mrad they develop cytotoxicity [125] that
has been attributed to the formation of glyoxal, $(CHO)_2$ [126]. In
addition, Schubert et al. [127] detected the formation of hydroxy-
alkyl peroxides by the interaction of hydrogen peroxide with carbonyl
compounds produced in the radiolytic breakdown of sucrose. For a
given radiation dose, these organic peroxides are more toxic to
bacteria than glyoxal [127]. Such reports of cytotoxicity may have
a bearing on the situation in sterilized soils according to the
nature and amount of the carbohydrates present. Glucose constituted
42-54% of the total sugars present in some mineral and organic soils
[128] and, furthermore, it is known that destruction of cellulose
during irradiation of peat with 10 Mrad results in an increase in
reducing sugars [129]. Cellulose in peat would be expected to
undergo less damage from radiation than cellulose in plant residues
and the soil biomass, because the higher percentage of lignin
associated with it would reduce the radiation scisson of cellulose
chains [130].

VI. GROSS BIOCHEMICAL CHANGES AND PLANT GROWTH IN IRRADIATED SOIL

A. Effect on Extractable Inorganic Ions and Organic Matter

Increases in the extractability of several elements following the treatment of soil over a wide range of radiation dose are summarized in Table 2. In some experiments in which autoclaving was directly compared with radiation it was found to raise the solubility of elements to a greater extent [41, 87]. Salonius et al. [117] reported 62% increase in conductivity of water extracts from autoclaved Lincoln clay, but only a 16% rise after irradiation. Exchangeable manganese showed a rise of seven-fold in steamed soil compared with two-fold increase after 2 Mrad, both relative to the concentration in untreated soil [87]. Several workers have found that irradiation of soils up to 3 Mrad caused no differences in pH [66, 87, 117, 132], but Boyer et al. [43] measured pH decreases of 0.25-0.6 units in three soils exposed to 4 Mrad.

Alteration in the composition of the soil solution was a better index of radiation damage to soil than measurement of changes in exchangeable nutrients, and lysis of dead microbial cells was believed to contribute the majority of carbon, nitrogen and some other elements set free [131]. An exception was manganese, for which the concentration in the soil solution after 5 Mrad seemed too large to have arisen from cell lysis. The enzymatic reduction of manganese dioxide may have been responsible, and this compound can serve as an electron acceptor for respiratory enzymes [137] that are capable of withstanding high-radiation treatments [87, 133, 138].

The extra ammonium extractable from soil shortly after irradiation with substerilizing doses is believed to result from death and lysis of the susceptible microbial population, with mineralization of organic N compounds by the surviving microflora and radioresistant enzymes [87, 133, 138]. A slow but continuous production of ammonium was detected over 12 weeks after irradiation

TABLE 2

Increases in Extractable Ions in Irradiated Soils

Soil	Dose (Mrad)	Elements showing increase[a]	Elements showing no increase[a]	Reference
Broad organic clay loam	0.0015	NO_3 [k]		[133]
Broad organic clay loam	0.2	NH_4 [k], NO_3 [k], org.C [s], org.P [s]	inorg.P [s]	[133]
Faringdon, Grove, Marl-borough and Ringwood light and heavy loams	0.2	NH_4 [k]		[87]
Broad and Icknield organic clay loams	0.6	NH_4 [k], NO_2 [w], NO_3 [k]		[93]
Zellwood peat	2	N [k], P [a]	Ca [a], K [a], Mg [a]	[134]
Faringdon, Grove, Marl-borough and Ringwood loams	2	Cu [a], Mn [a]	K [a], Zn [a]	[65]
Faringdon and Grove loams	2	NH_4 [k], NO_3 [k]	S [k]	[65]

(continued)

Table 2, continued

Soil	Dose (Mrad)	Elements showing increase[a]	Elements showing no increase[a]	Reference
Ahuriri silt and Twyford sandy loams	2.5		Mo (s)	[135]
Lincoln clay	3	B (w), Ca (w), Cu (w), Mg (w), Mn (w)	K (w), Na (w)	[117]
Everglades muck	3	S (m)		[41]
Faringdon sandy and Grove clay loams	5	org.C (s), Cu (s), org.Mg (s), Mn (s), org.N (s), NH$_4$ (s), org.P (s)	Ca (s), Cl (s), Fe (s), Na (s), Zn (s)	[131]
Broad organic clay loam	12	NO$_2$ (w)		[136]
Faringdon sandy and Grove clay loams	60	NH$_4$ (k)		[65]

[a]Extractants used: (a) ammonium acetate solution, (k) potassium chloride solution, (m) Morgan's solution, (s) soil solution, and (w) water.

of Faringdon and Grove soils but no gaseous ammonia was released
from soil treated with 2.5 Mrad [65]. The enzymic production of
ammonium in soil irradiated sufficiently to inactivate all microbes
has been observed [7, 106]. A further rise in dose leads to greater
inhibition of enzymes, but massive irradiation of soil up to 60
Mrad can still result in the formation of ammonium amounting to 76
ppm NH_4-N by 24 hr after exposure [65]. The deamination of organic
N compounds is almost certainly responsible, and "G" values of
0.060 and 0.061 have been calculated for the ammonium-N extracted
from two soils given 30 Mrad [65]. (G is the number of molecules
of product formed per 100 eV of energy absorbed, and 1 rad corresponds to the deposition of 6.24×10^{13} eV/g of target material.)
When the same pair of soils were boiled with ethanol to inhibit
enzymes before the application of 5 Mrad, similar G values were obtained to those after 30 Mrad: the chemical action of radiation
was therefore responsible for the increase in ammonium. Soil moisture content at the time of irradiation and during subsequent incubation has an important influence on the amount of ammonium extracted. When air-dry soils were treated up to 2.5 Mrad and incubated without rewetting, no extra ammonium was formed, but if moistened with sterile water immediately after γ treatment they released
as much ammonium as soils irradiated in a moist state [65]. Such
behavior is consistent with inhibition of enzymes under dry conditions, and Skujins and McLaren [139] have shown that the activity
of urease in a Dublin clay loam is negligible below 80% relative
atmospheric humidity.

The amount of total mineral N extractable from soil that has
been incubated for a short period after irradiation to permit mineralization of organic N is important regarding the question of γ-
degradation of humus. The quantities of mineral-N removed from
several soils after 0.25-2.5 Mrad appear in Fig. 3 [140], and other
work has shown a similar relationship [32]. The distinct plateau
effect suggests that up to 2.5 Mrad the deamination of organic matter is slight relative to the release of ammonium from the dead
biomass, and the large rise in total mineral N between 0.25 and 2.5

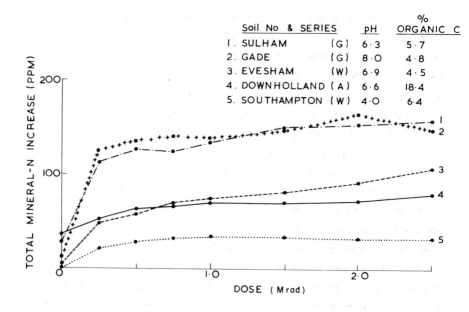

Fig. 3. Increases in the concentrations of total mineral nitrogen in arable (A), grassland (G), and woodland (W) soils incubated for 7 days (25°C) after irradiation [140].

Mrad corresponds to the dose region at which one would anticipate severe damage to the biomass.

Soils fallowed in the open for 1 year after exposure to 2.5 Mrad were no different from nonirradiated samples with respect to organic C and N extractable with 0.5 N NaOH, or the proportions of these elements in the fulvic and humic fractions [131]. Peat is not easily degraded until doses approaching 10 Mrad are used [129], and the irradiation of soil organic matter extracts indicated that about 4% of the ammonium produced in Grove clay loam by 5 Mrad comes from degradation of humus: the fulvic fraction contributed most of this ammonium [131]. In addition, Jenkinson [141] concluded that the application of 4 Mrad to a calcareous soil (2.4% organic C) rendered small additional amounts of organic matter decomposable by processes other than killing of the soil biomass.

The separation of humic and fulvic fractions of organic matter from seven soils that had received 4 Mrad revealed increases of 40-100% in fulvic acids (extractable with sodium pyrophosphate) but little additional humic component [43]. It is noteworthy that enrichment of one soil with glucose prior to irradiation led to a rise of at least three-fold in extractable fulvic acids [43], almost certainly from damage to the extra biomass that developed during incubation before γ treatment. Hering [142] used 2.5 Mrad to sterilize oak-leaf litter before inoculating it with several species of fungi to examine decomposition rates: sterilization by heat (20 lb/in^2 for 1 hr) was less suitable because it induced faster breakdown of inoculated litter.

B. Oxidation of Ammonium

The treatment of several fresh soils with 0.05-0.2 Mrad resulted in faster accumulation of nitrate during 1 week of incubation, and nitrification was stimulated in one soil (Broad series) by only 0.0015 Mrad [133]. A small amount of extra nitrate appeared in Arredondo fine sand exposed to 0.06 Mrad [143]. Experiments over a wider dose range, up to 3 Mrad, revealed that oxidation of ammonium to nitrate on incubation usually reached a maximum after 0.2-0.8 Mrad, greatly exceeding the rate of nitrification in nonirradiated soil [32]. Further extension of the dose from 0.8 to 3 Mrad gradually inhibited nitrification with a corresponding rise in ammonium concentration, and this typical relationship is presented in Fig. 4 for Grove clay loam. Although the proliferation of nitrifying bacteria in perfused soil was prevented by 0.1-0.2 Mrad, Fig. 1 shows that the organisms could still oxidize a limited quantity of ammonium to nitrate: on a quantitative basis, this accounted for the increase in nitrate recorded in nonperfused soil that had been similarly irradiated and incubated for 7 days [32]. The radioresistance of enzymes involved in the pathway of ammonium oxidation to nitrate is thus demonstrated.

In some experiments, soils irradiated with substerilizing doses have failed to show greater accumulation of nitrate [66, 84, 98].

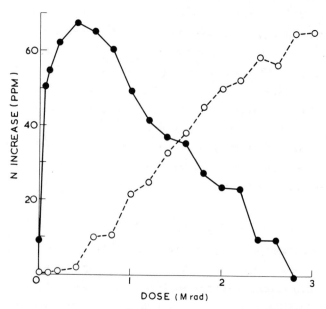

Fig. 4. Increases in the concentrations of ammonium (O- - -O) and nitrate (●——●) in Grove clay loam incubated for 7 days (25°C) after irradiation [32].

In one instance ammonium was added before irradiation [98]; therefore, the detection of a nitrification response to the relatively small amount of radiation-released ammonium is unlikely. This is borne out by the results obtained under ammonium perfusion, which appear in Fig. 1 [32]. In other cases the poor increase in nitrification may result from a low initial population of nitrifying bacteria either occurring naturally or induced by air-drying soil prior to irradiation [66, 98]. To resolve these problems, nitrate was analyzed in ten soils irradiated in fresh condition (at 45-65% WHC) with 0.25-2.5 Mrad and then incubated for 1 week [75]. Eight of the soils showed an increase in nitrate concentration, up to 14-fold in excess of the amount formed by nonirradiated but incubated controls. In five of the soils, 2.5 Mrad did not reduce nitrification

below that of control samples. The two irradiated soils that failed
to nitrify at a faster rate were from acid woodland (pH 3.1) and
grassland (pH 5.1), although ammonium accumulated in both and was
not limiting formation of nitrate. The rate of ammonium oxidation
in soils treated with 0.25 Mrad was well correlated with the ability
of nonirradiated soils to nitrify ammonium sulfate [75], and the
original population or activity of nitrifying organisms appeared
to be the main factor limiting nitrate formation after γ treatment
in the region stated. The application of 0.8 Mrad to soil at moistures above 75% WHC resulted in markedly less accumulation of nitrate on incubation compared with less saturated samples, and biological reduction of nitrate to oxides of nitrogen and gaseous
nitrogen was considered responsible [31].

C. Reduction of Nitrate and Nitrite

According to the radiation dose applied to certain soils, biological or purely chemical (radiolytic) reduction of nitrate can
lead to temporary accumulation of nitrite [93, 136]. The biological
pathway is inhibited above 4 Mrad but radiolytic reduction becomes
important [144]. It is most convenient to discuss these different
mechanisms of reduction with reference to the separate dose regions:

1. Treatment up to 4 Mrad

Cawse and Crawford [93] noticed that nitrite and nitrate
accumulated rapidly when two organic calcareous soils were irradiated with 0.6-2.5 Mrad. A typical response from Broad series, kept
at 25°C for 7 days after irradiation, is shown in Fig. 5 [144]. No
nitrite formed in Faringdon sandy loam (Fyfield series) and the
sharp rise in nitrite concentration in Broad soil could be prevented
by autoclaving the sample before irradiation. Two principal mechanisms were considered responsible for the formation of nitrite:
first, the inhibition of *Nitrobacter*-induced nitrite oxidation, or
second, the reduction of nitrate [93]. The latter mechanism was
favored because leaching irradiated (0.6-0.8 Mrad) soils with water

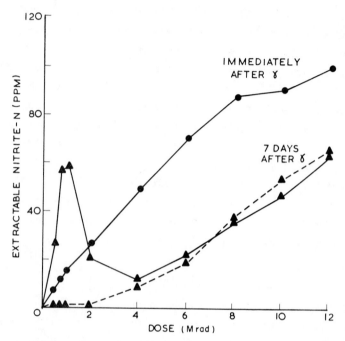

FIG. 5. Nitrite present immediately and 7 days after irradiation of Broad clay loam (————) and Faringdon (Fyfield series) sandy loam (- - -). Soils contained 200 ppm nitrate-N when exposed to γ radiation [144].

or ammonium sulfate solution prevented accumulation of nitrite, but leaching with potassium nitrate solution did not [93]. Nitrate labeled with ^{15}N was reduced to $^{15}NO_2$ in calcareous soils treated with 0.5-0.8 Mrad, with oxidation of ammonium making a smaller contribution to the nitrite pool [145]. Nitrate reduction was thought to be carried out by radioresistant enzyme systems of denitrifying bacteria, such as *Bacillus* and *Pseudomonas* spp., which are unable to proliferate, and it is known that these organisms are favored by soils of high pH [146, 147].

When ten soils were given 0.25-2.5 Mrad the optimum region for formation of nitrite was between 0.5 and 0.8 Mrad and by 24 hr after irradiation, nitrite (26-60 ppm NO_2-N) appeared only in those soils

containing carbonate [148]. The accumulation and persistence of
nitrite was poor in soils with pH<7 and was not improved by the
addition of calcium carbonate before irradiation [31]. This suggests
that these soils do not possess enzymes capable of reducing nitrate
after irradiation and that rapid decomposition of nitrite at low
pH [149] is not responsible for its failure to accumulate.

The accumulation of nitrite in 15 fresh soils over 64 hr after
treatment in the region of 0.75 Mrad was influenced by incubation
temperature, moisture, soil aggregate size, and storage conditions
before irradiation, all factors known to affect microbial activity
[31]. Incubation of irradiated Broad and Icknield soils at temperatures from 0° to 40°C increased the amount of nitrite formed, but
higher temperatures prevented accumulation. Raising the soil moisture from air-dry to 75% WHC caused more nitrite and nitrate to
appear, although the concentrations of both anions fell sharply at
higher moisture, presumably from the activity of reducing enzymes.
As aggregate diameter increased from <0.5 to 3 mm less nitrite was
formed: Nommik [150] reported a fall in denitrification rate in
nonirradiated soil at saturation point when aggregate size increased
to 2-4 mm, and improved oxygen diffusion was considered responsible.
Air-drying soil before irradiation with 0.8 Mrad decreased the subsequent accumulation of nitrite under moist conditions [31]. The
rates of biological nitrate reduction in nonirradiated soils that
were waterlogged (with simultaneous addition of glucose and nitrate)
were closely correlated with the ability of the soils to accumulate
nitrite after exposure to 0.75 Mrad at medium moisture levels [31].

Gas chromatography of the soil atmosphere in irradiated (0.75
Mrad) Broad, Broadmoor, and Icknield calcareous clay loams proved
that nitrite was being formed at oxygen concentrations from 12.5 to
16.1% by volume (at 20°C, 760 mm Hg) [115]. Campbell and Lees [151]
have discussed the activity of nitrate reductase in relation to
oxygen tension and Hauck [152] has suggested that the enzymic reduction of nitrate may occur in anaerobic micro-zones at the center
of soil aggregates under apparently aerobic conditions. Several

workers have recorded losses of nitrate in soils kept well aerated [153-155].

Skyring and Thompson [88] stated that 8% of nitrate in 5 Mrad treated black earth was reduced to nitrite, but only when moist rather than dry soil was irradiated. In moist soil that had received 2.5 Mrad, Jenkinson and Powlson [110] detected 4 ppm NO_2-N and a corresponding decrease in nitrate. In both of these papers it is uncertain whether biological or radiolytic reduction of soil nitrate was responsible for the appearance of nitrite.

Nitrite formed by biological activity after irradiation of soil with 0.5-0.8 Mrad is partially removed during periods of incubation longer than 3-7 days [144, 148]. For example, the peak of almost 60 ppm NO_2-N in irradiated Broad soil (Fig. 5) had fallen to 46 ppm after 21 days, and in Icknield loam there was 59 ppm NO_2-N at 7 days compared with 19 ppm after 21 days [144]: experiments on the loss of $^{15}NO_2$ tracer added to Icknield soil treated with 0.8 Mrad showed that nearly 60% was oxidized to nitrate after 14 days of incubation, with 1% assimilated by microorganisms. Fixation to organic matter (i.e., nonextractable with water) and volatilization, mainly as N_2O and N_2, accounted for the remaining deficit [114]. It is therefore evident that the balance in the activities of nitrifying and denitrifying bacteria in irradiated soil is a major factor influencing the initial rise and subsequent fall in nitrite. It has recently been shown that air-drying Icknield soil for 2 days before rewetting to 79% WHC also leads to rapid accumulation of nitrite over 2 days of incubation, and $^{15}NO_2$ was produced from ^{15}N-labeled nitrate [156]. When three other soils were air-dried and rewetted, nitrite did not appear in the two that failed to produce nitrite after exposure to 0.75 Mrad. One possibility is that the release of carbohydrates from damaged microorganisms following both air-drying and irradiation, with subsequent oxidative metabolism, resulted in an increase in the extent of anaerobic microzones, which, in turn, stimulated nitrate reductase.

2. Treatment up to 12 Mrad

In 1969, Singh et al. [157] reported a loss of added nitrate when some Hawaiian soils were treated with 3 Mrad, and believed that certain chemical reactions were responsible instead of biological denitrification. In a subsequent study [138] several soils given 5 Mrad lost even more nitrate: nitrite was not analyzed separately, although it appears that the steam-distillation procedure used to recover nitrate would have included any nitrite that remained. Cawse and Cornfield [144] examined the formation and decomposition of nitrite in four soils of pH 7.4-8.0, adjusted to 200 ppm NO_3-N and 60-80% WHC before receiving 0.5-12 Mrad. In the selection of these soils, Broad and Icknield types were chosen for their ability to form nitrite after 0.75 Mrad, and Faringdon and Grove were selected for their inability to do so [31]. Results from the representative Broad and Faringdon soils appear in Fig. 5. Immediately after irradiation all four soils showed increasing accumulation of nitrite in relation to dose, typified by Broad loam, and the rise in nitrite was correlated with a decrease in nitrate [144]. Autoclaving soils before irradiation with 12 Mrad did not prevent nitrite from forming and the γ radiolysis of nitrate was believed responsible [144]. Irradiation of soils at lower moistures decreased the yield of nitrite, and if their nitrate content was restricted to 40 ppm NO_3-N, less than 25 ppm NO_2-N appeared immediately after 12 Mrad [144].

The formation of nitrite in heavily irradiated nitrate solutions has been previously reported [158, 159], and Pikaev [160] discussed the mechanism of nitrate radiolysis. However, many aspects of this reduction pathway remain uncertain and are under investigation [161]. When a solution of potassium nitrate (200 ppm NO_3-N) received 2 Mrad, 6.9 ppm NO_2-N was found immediately afterward giving a G value of 0.24; the yield of nitrite was slightly better in irradiated soil solutions but was greater still in the complete soil system [144]. Clay and organic matter may influence free-radical reactions, and experiments on irradiation of model systems are needed to resolve this point.

Analysis of soils at 7 days after 12 Mrad showed that water-extractable nitrite had decreased in all of them, as indicated in Fig. 5 for Broad and Faringdon loams [144]. By 21 days after irradiation, 79% of the nitrite initially formed could not be recovered. Fixation of nitrite to organic matter and volatilization as nitric oxide, nitrogen dioxide, and nitrogen were almost equally responsible for this decline, and the rate of loss was increased by the addition of lignin or catechol [144]. The formation of nitrosophenols with nitrite and the reaction of such compounds with nitrous acid to give gaseous loss of nitrogen may be responsible [162]. It is important to note (Fig. 5) that by 7 days after irradiation with up to 2 Mrad, the amount of radiolytically produced nitrite present is negligible in relation to nitrite that originates in certain soils by biological reduction of nitrate.

Irradiation of soil with 12 Mrad has been suggested as an alternative to heat sterilization for studies on nitrite fixation to organic matter and volatilization as nitrogen oxides and gaseous nitrogen [144]. Nitrite is formed in situ by radiolytic reduction of nitrate, which obviates the need to disturb the sample by adding sterile nitrite solution after heat treatment. If accumulation of nitrite is a problem in experiments that require heavily irradiated soil, several steps can be taken to prevent it. Soil could be leached before γ treatment to remove nitrate, irradiated under air-dry condition, or left for approximately 1 month after irradiation to allow nitrite to decompose. In addition, the choice of soil with a low pH would encourage rapid removal of nitrite.

Investigations on the formation of nitrite by calcareous soils treated with 0.25-12 Mrad have thus demonstrated that γ-induced changes in soils can arise from chemical (radiolytic) reactions as well as biological processes. Caution is therefore necessary in the interpretation of microbial and enzyme inactivation experiments, particularly when 4 Mrad is exceeded, and this aspect is again mentioned in connection with the release of CO_2 from irradiated Icknield loam. It is notable that biological reduction of nitrate

to nitrite after γ exposure of soil can take place under apparently
good aeration, simultaneously with extensive oxidation of ammonium
to nitrate [31, 145]; if nitrite does accumulate by biological activity in irradiated soil, the possibility that it originates from
oxidation of ammonium should be borne in mind unless reduction of
added $^{15}NO_3$ tracer proves otherwise.

Present knowledge of nitrite formation and decomposition in
soil treated with γ rays now enables a better approach when radiation is used for research on the nitrogen cycle. However, further
characterization of soils that can biologically reduce nitrate after
moderate irradiation is needed, especially in relation to the nature
and density of their denitrifying and total microflora.

D. Release of Carbon Dioxide

The application of acute substerilizing and sterilizing doses
of radiation to soils has resulted in widely different observations
on the subsequent release of CO_2. Particular attention should be
paid to differences in experimental technique that may affect,
first, the radiosensitivity and hence the extent of mineralization
of killed organisms, and second, the recolonizing ability of microbes.
For example, although treatment of fresh soils with 0.2 Mrad stimulated the release of CO_2 [133], irradiation of air-dry Rifle peat
with 0.25 Mrad did not affect CO_2 evolution over 54 days of incubation in a remoistened state [66]. Air-drying is, in itself,
known to result in the death of many organisms and an intensive CO_2
flush when soil is rewetted [39, 47], and the extra death and mineralization of radiation-damaged cells may be small at substerilizing
doses. Evidence for the microbial origin of CO_2 was obtained by
Grossbard [163], wherein the addition of a dense suspension of irradiated cells to nonsterile soil increased the CO_2 output.

Popenoe and Eno [98] found that addition of glucose to fresh
soil treated with 0.02 and 0.2 Mrad caused an initial depression
in CO_2 production relative to the glucose-stimulated control.
Therefore, it appears that the presence of ample substrate for

microorganisms may obscure the stimulatory effect of a relatively
small amount of radiation-released substrate on the surviving cells.
In the same way, perfusion experiments described earlier in this
review have shown (Fig. 1) that addition of ammonium to irradiated
soil prevents the response of nitrifying bacteria to radiation-
released ammonium from being observed [32].

Recent examination of the CO_2 release from fresh Icknield loam
(60% WHC) after exposure over the range 0.025-10 Mrad proved that
all levels of radiation stimulated the evolution of CO_2 compared
with nonirradiated soil [30]. Most rapid evolution of the gas
occurred in the first day after γ treatment and a significant amount
of CO_2 appeared during irradiation, although 10 days later both
irradiated and untreated samples released CO_2 at similar rates.
Several workers who have used 2.5 Mrad for soil sterilization have
noted that the postirradiation release of CO_2 falls rapidly over 2
days to approach the normal control respiration [16, 110, 120].
Cores of litter-soil given 2.5 Mrad and returned to the field showed
a similar respiration rate to nonirradiated samples [120].

Peterson [29] reported that the application of 3.3 Mrad to
Chalmers clay loam reduced O_2 uptake by only 38%, and dead microbial cells were considered to be still capable of respiration. It
was proved that this respiratory activity could be stimulated by
the addition of glucose or depressed by mercuric chloride. Meisel
et al. [164] had previously shown that inactivation of yeast in
pure culture by 1 Mrad reduced the respiration rate by less than
50%. Roberge [165] studied the activity of respiratory enzymes in
moist (60% WHC) spruce humus sterilized with heat or 1.1-4.4 Mrad
γ radiation and concluded that respiratory enzymes were responsible
for most of the O_2 uptake recorded. Biological O_2 uptake was indicated, because the addition of sodium azide to irradiated humus
reduced respiration whereas urea slightly increased it. After a
higher exposure of 4.4 Mrad O_2 uptake was almost completely inhibited [165], in agreement with the radiation dose needed to inactivate
urease in this humus [102].

It is by no means certain that the CO_2 released from irradiated soil is always derived from enzyme activity, a second possibility being the decarboxylation of organic compounds by radiation [63, 166]. Following treatment of soil (2.4% organic C) with 4 Mrad, Jenkinson [141] detected a small release of CO_2 from the nonbiomass section of soil organic matter, although the majority of this gas originated from the killed biomass. Irradiation of Icknield loam (10.2% organic C), selected because of its high natural rate of respiration, indicated that formation of CO_2 by enzymes was >95% predominant up to 2 Mrad. However, when the soil was given 10 Mrad, 45% of the CO_2 evolved came from the γ-decarboxylation of organic matter [30]. Treatment of Icknield soil over the range 4-10 Mrad gave little extra yield of CO_2, and it was thought that the inhibition of CO_2 production by biological means may have been compensated by more γ-induced decarboxylation [30]. It is interesting to note that in this event, the pattern of behavior would be similar to formation of nitrite in soil irradiated up to 12 Mrad: the initial yield results from biological activity, but as the dose is raised this mechanism becomes inhibited and is replaced by increasing radiolytic reduction of nitrate to nitrite (Fig. 5). Further research is required to accurately define pathways leading to the evolution of CO_2 from soils treated with 2.5 Mrad or more.

The delivery of 0.3 Mrad (accumulated air dose) to γ-field soil over 3 months produced no microbial or respiratory response [167]. The continuous influx of organisms was believed to compensate for radiation damage and natural variations in moisture greatly affected the respiration rate, probably obscuring the γ effect. Clark and Coleman [16] reported that oak litter soil exposed in a γ-field to 0.8 Mrad over 6 weeks showed slightly improved respiration when incubated for 4 weeks. A soil microbial flora established in nutrient-enriched sand respired less CO_2 when 0.5 mCi of ^{32}P was added and qualitative differences in the microflora were found: these effects were not observed when a silt loam was similarly treated [168].

If the extra evolution of CO_2 is a problem in certain experiments,

use of irradiated soils can be delayed for a week to allow high concentrations of CO_2 to diffuse out. Alternatively, the gas could be
displaced with air either during or immediately after irradiation,
and the exposure of soil contained in acrylic plastic columns [15]
would facilitate this operation.

E. Degradation of Compounds
Added to Soil before and after Irradiation

Because irradiation of soil is a valuable cold-sterilization
technique that has little effect on humus, it is very applicable
to studies on the adsorption of various compounds to clay and organic
matter and their inactivation by microorganisms. These materials
should be added to soil after irradiation unless good evidence exists
for their radiation stability. Van Groenewoud [169] established
that 2.5 Mrad had practically no effect on the action of five anti-
biotics on *Staphylococcus* and suggested that γ sterilization and
inoculation of soil was ideal for research on the antagonism of
microorganisms in their natural media. However, work with other
compounds has frequently shown breakdown when irradiated, and Getzin
and Rosefield [170] found that the insecticides diazinon, malathion,
and zinophos were partly decomposed by 4 Mrad. According to Lippold
et al. [171], solutions of pesticides treated with 5 Mrad showed
the following reduction in concentration: DDT, 75%; malathion, 71%;
binapicryl, 58%; endosulphan, 45%; and parathion, 27%. Numerous
degradation products appeared,and the authors concluded that irradiation may similate natural processes of degradation in soil and
provide a useful means of examining breakdown pathways of compounds.
The herbicide atrazine was substantially degraded when exposed to
2.5 Mrad in moist soil [172].

With regard to reports on the degradation of compounds added to
radiation-sterilized soil, malathion, dichlorvos, ciodrin, and
mevinphos were decomposed at a faster rate in irradiated Chehalis
clay loam compared with autoclaved soil, and a heat-labile substance
that accelerated malathion degradation was extracted from both

γ-sterilized and nonsterile soil [170]. Parathion, dimethoate, and
diazinon were decomposed at similar rates in either heat or γ-
sterilized soil [170]. Graham-Bryce [173] exposed Broadbalk heavy
loam and Woburn sandy loam to 2.5 Mrad and obtained an improvement
in the recovery of disulfoton: microbial decomposition of this
insecticide in nonsterile soils prevented a correct assessment of
adsorption, which was closely related to the organic matter content.
Bowmer [172] examined the breakdown of ^{14}C-ring-labeled atrazine
in 2.5 Mrad-treated soils: in Fladbury clay top-soil (pH 6.4, 10%
organic C) the persistence of atrazine was not improved by irradiation, but in a sandy loam (pH 4.2, 1.9% organic C) it caused the
rate constant of atrazine disappearance to be reduced by up to 50%.
It was finally concluded that soil texture and pH were the primary
factors influencing atrazine disappearance rather than biological
degradation, and other work has shown that deactivation of atrazine
occurs readily in autoclaved soils [174, 175].

Nomenclature of Compounds

Atrazine	2-Chloro-4-ethylamino-6-isopropylamino-s-triazine
Binapicryl	3-Methylcrotonic acid 2-sec-butyl-4, 6-dinitrophenyl ester
Ciodrin	α-Methylbenzyl-3-hydroxycrotonate dimethyl phosphate
DDT	1,1,1-Trichloro-2,2-bis(p-chlorophenyl)ethane
Diazinon	Phosphorothioic acid O,O-diethyl-O-(2-isopropyl-6-methyl-4-pyrimidinyl)ester
Dichlorvos	Phosphonic acid 2,2-dichlorovinyl dimethyl ester
Dimethoate	O,O-Dimethyl-S-methylcarbamoylmethylphosphoro-dithioate
Disulfoton	Diethyl-s-[2-(ethylthio)ethyl]phosphorothiolothionate
Endosulfan	1,4,5,6,7,7-Hexachloro-5-norbornene-2,3-dimethanol cyclic sulfite
Malathion	S-(1,2-dicarbethoxyethyl) O,O-dimethyldithiophosphate
Mevinphos	2-Carbomethoxy-1-methylvinyldimethyl phosphate
Parathion	O,O-Diethyl-O-p-nitrophenyl phosphorothioate
Zinophos	O,O-Diethyl-O-2-pyrazinyl phosphorothioate

F. Plant Growth in Irradiated Soil

Variable increases in plant growth have often been observed in
soils previously irradiated with substerilizing or sterilizing
doses, and some results are summarized in Table 3. Fertilizer was
not added to soils in the experiments quoted. In the trials using
steam or chemicals in addition to radiation, increases in plant yield
are calculated in relation to the tissues harvested from nonsterile
soil. Superior development of plants in irradiated soil compared
with that sterilized by other techniques has been attributed to
inhibition of growth by residues of formaldehyde and methyl bromide
[121], and unidentified toxic substances after steaming [87, 177].

The extra yield of lettuce in Grove soil exposed to only 0.005
Mrad [65] corresponds with a rise in soil nitrate concentration [133].
Lettuce removed two to four times more nitrogen from soils treated
with 0.02-0.2 Mrad than from nonirradiated ones [65], and this was
related to the γ-induced rise in exchangeable nitrogen. When Grove
clay loam was given 2 Mrad and fallowed in the open for 1 year before sowing with lettuce, the yield was 64% greater than that obtained from just fallowed soil [178]. Soil irradiation with 0.2
Mrad increased the concentrations of manganese, phosphorus, and
nitrogen in lettuce and raised the total uptake of these elements
[87]. Eno and Popenoe [134] found improved uptake of nitrogen and
phosphorus by oats grown in Scranton fine sand and Zellwood peat
given 0.5 and 1 Mrad. The large increase in lettuce yield from γ-
treated Marlborough soil (Table 3) is associated with extremely poor
growth in normal soil, possibly limited by available phosphorus:
the dry weight of tissue harvested from irradiated soil only amounted
to 50% of the quantity produced by Faringdon and Grove soils exposed
to 2 Mrad [87].

Growth of Blackbutt (*Eucalyptus pilularis*) improved after the
application of 2.5 Mrad to an Australian soil, and it was thought
that a fungus that inhibited root development had been eliminated
[179]. Culture of wheat in 2.5 Mrad-treated and reinoculated

TABLE 3

Plant Growth in Irradiated Soils
and Soils Sterilized by Heat or Chemicals

Dose (Mrad)	Soil[a]	Test plant	% Increase in yield relative to untreated soil	Reference
0.005	G	Lettuce	17	[65]
0.01	G	Lettuce	28	[65]
0.02	G	Lettuce	67	[65]
0.02	F	Lettuce	55	[65]
0.03	C	Rye	20	[176]
0.08	G	Lettuce	54	[65]
0.08	F	Lettuce	99	[65]
0.2	G	Lettuce	66	[65]
	G	Flax[b]	91	[65]
0.2	G	Lettuce	69	[87]
Steam	G	Lettuce	- 3	[87]
0.2	F	Lettuce	52	[65]
0.2	F	Lettuce	24	[87]
Steam	F	Lettuce	56	[87]
1	M	Lettuce	269	[87]
2	M	Lettuce	301	[87]
Steam	M	Lettuce	41	[87]
2	F	Lettuce	36	[87]
2	G	Lettuce	42	[87]
	G	Flax[b]	118	[65]
2.5	W	Ryegrass	133	[121]
MeBr	W	Ryegrass	73	[121]
Formalin	W	Ryegrass	67	[121]
2.5	W	Wheat	108	[121]
MeBr	W	Wheat	30	[121]
Formalin	W	Wheat	47	[121]
4	A	Ryegrass	81	[43]
4	C	Ryegrass	30	[176]
42	L	Ryegrass	30	[176]

[a]Soil Nomenclature: (A) Antony brown earth, (C) Cecil sandy loam, (F) Faringdon (Fyfield series) sandy loam, (G) Grove clay loam, (L) Lakeland sand, (M) Marlborough clay loam, and (W) Woburn sandy loam.

[b]As following crop to lettuce; soil only irradiated once.

red-brown earth gave three times the grain yield obtained from untreated soil, and the high asparagine content of plants growing on
the irradiated medium indicated an ammonium N nutrition from the
extra nitrogen released [89]. Plants from nonirradiated soil contained more glutamine, glycine, and α-alanine, but the total quantity
of amino acids per gram fresh tissue did not differ between treatments.

Ulrich et al. [180] leached soluble nitrogenous compounds and
other materials from electron-sterilized soil before experimental
use and found that growth of tomato was identical to that in a nonsterile sample. The irradiated medium was used to examine the
utilization of added protein by plants in the absence of protease
activity. Loutit et al. [181] treated soil with 2.5 Mrad for studies
on the effect of microorganisms on molybdenum uptake by radish and
later reviewed the importance of microbial and nonmicrobial factors
on plant molybdenum content [182]. Radish cultured in γ-sterilized
Ahuriri silt loam contained more molybdenum in the leaves and stems
than plants from non-sterile or sterile samples inoculated with
rhizosphere organisms: sodium molybdate was added to all treatments
[135]. Irradiation of white mustard plants while they were growing
in Broad clay loam showed that 0.005 Mrad depressed growth but increased the phosphorus content of the tissues at harvest [183].
Lysis and mineralization of the killed microflora are not necessarily
responsible for this effect, because plant uptake of phosphorus can
occur by a radioresistant translocatory pathway [184].

The interaction of fertilizer and irradiation of soil on plant
growth has been examined. The yield of barley leaf tissue was improved by 0.2 Mrad regardless of whether ammonium nitrate was applied [65], and ryegrass behaved similarly when four types of nitrogen and phosphorus fertilizer were added to soil exposed to 4
Mrad [43]. Jenkinson et al. [121] used "N values," calculated from
the uptake of fertilizer nitrogen by wheat and ryegrass, to indicate
the value of γ-released nitrogen to these plants. An improved
growth of wheat in irradiated soil, either with or without calcium

nitrate fertilizer, was attributed entirely to the extra nitrogen
mineralized. With ryegrass, there was evidence to suggest that a
factor in addition to radiation had partly contributed to better
growth, but there was no indication that the control of soil-borne
plant pathogens by radiation had improved plant development [121].
In an experiment on irradiation of Cecil sandy loam and Lakeland
sand adjusted to six nitrogen, phosphorus, and potassium fertility
levels, Cummins and McCreery [176] found that uptake of calcium,
magnesium, phosphorus, potassium, and sodium by rye was unaffected
except where influenced by an increased yield from γ-released mineral nitrogen.

With regard to inhibition of plant growth, Boyer et al. [43]
reported 18% reduction in clover herbage from 4 Mrad-treated Antony
brown earth, with a similar result after the application of nitrogen
and phosphorus fertilizer together with radiation. It was shown that
root nodules were completely absent after the elimination of *Rhizobium*. Stanovick et al. [86] described poor and chlorotic growth of
vernal alfalfa in γ-sterilized soil, and *Rhizobium meliloti* was
proved absent. Bowen and Rovira [185] observed a slight decrease in
growth of subterranean clover in soil treated with 2.5 Mrad, and root
hairs were short and sparse. More severe inhibition occurred following
sterilization by heat or propylene oxide. In subsequent work Rovira
and Bowen [186] showed that the addition of an eluate from heat-
sterilized Urrbrae red-brown earth to clover growing in an irradiated
sample resulted in inhibition of primary root growth, but detoxica-
tion by microorganisms could be achieved by applying a soil inoculum
to the sterile sample. The initial formation of organic materials
by heat treatment was thought more likely to result in toxicity
than were changes in pH, soluble manganese and ammonia. Phytotox-
icity in urea-fertilized soils has been partly attributed to accumulation of nitrite [187], and solution culture of plants has been
used to demonstrate nitrite toxicity [188, 189]. Nevertheless, the
formation of 35 ppm NO_2-N in Broad loam after treatment with 0.8 Mrad
did not prevent an increased yield of lettuce from being recorded [93].

Table 3 shows that lettuce grown in Faringdon soil given 0.2
Mrad gave considerable differences in growth response from one experiment [87] to another [65], at an interval of 2 years. Responsible factors may be variation in the degradation rate of an inhibitory compound between the irradiated soils, or a change in total
microbial population that affects the quantity of γ-released nitrogen available to plants. It is known that field fumigation with
formalin or methyl bromide reduces the biomass sufficiently to decrease the amount of nitrogen mineralized after exposure of soil
to 2.5 Mrad, even when irradiation is performed several years later
[110]. This finding agrees with a gradual decline in crop response
to repeated soil fumigation [190].

G. Other Effects

1. Sulfur Oxidation

Microbial oxidation of elemental sulfur to sulfate in Arredondo
fine sand was gradually inhibited by increasing radiation up to 2
Mrad, and two groups of organisms appeared to be damaged [98]. The
most sensitive group was severely affected by 0.016 Mrad and may
have included filamentous fungi or actinomycetes [137]. The greater
resistance of the second group, which were still active after 0.06
Mrad, seems to indicate that they included chemoautotrophic bacteria
of the genus *Thiobacillus* [137].

2. Soil Aggregation

The stability of aggregates in air-dried Lincoln clay was reduced by 6% following treatment with 3 Mrad [117]. However, Griffiths
and Burns [191] reported that the stability of natural aggregates in
four soils was unaffected by 2.5 Mrad, although it was reduced by
irradiation of synthetic aggregates prepared with polysaccharides
isolated from the soil yeast *Lipomyces starkeyi*. The authors suggested that irradiation of soil could be used to assess the relative
importance of microbial polysaccharides in aggregation. Boyer et
al. [43] found that the percentage of water-stable aggregates between

0.2 and 2 mm was decreased by 4 Mrad, especially when the microflora
of one soil was supplied with glucose to give a measured improvement
in stability before irradiation. The rate of water percolation
through soil was reduced after 2.5 Mrad, and γ-treated soil dried
out at a slower rate than an untreated sample [65]; however, plant
growth was greatly improved and any deleterious γ effect on aggregation that affected root development appeared to have been out-
weighed by the increase in available nutrients.

3. Nitrogen Fixation (Nonsymbiotic)

The survival of *Azotobacter* has been examined by inoculation
of culture media with irradiated soil [17, 18], but little information is available concerning the effect of radiation on nitrogen
gains in soils that are naturally capable of significant nonsymbiotic nitrogen fixation. A recent study of this aspect by Sheldon
[192] involved treatment of Icknield silty clay loam, saturated
with water to a depth of 3 mm, with 0.125-4 Mrad. Nitrate was
leached from the soil before irradiation to avoid an inhibitory
effect of this anion on nitrogen fixation from the atmosphere [193-
195], and also to reduce the chance of nitrogen loss by decomposition
of radiolytically formed nitrite [144]. The normal increase in total
nitrogen in saturated Icknield soil over 6 weeks at 25°C was slightly
inhibited by 0.25 Mrad but was not completely inhibited until 0.75-
1 Mrad had been initially applied [192]. Such a dose is considerably higher than the 0.1-0.2 Mrad required to prevent cell division
in *Azotobacter* [17, 62] and *Clostridium* [54], and the continued activity of nitrogenase [151] is indicated. Some *Azotobacter* may
have survived waterlogging, but the anaerobic *Clostridia* were considered [192], from other evidence [196], to be the main bacteria
responsible for nitrogen fixation under these conditions. It should
be possible to establish cultures of *Azotobacter* and *Clostridium*
separately in γ-sterilized soil for precise experiments on the response of their nitrogen-fixing activity to irradiation in a soil
environment.

4. Release of Ethylene

Although studies by Abeles et al. [197] with two Maryland soils at field capacity indicate that microbial degradation removes ethylene and other hydrocarbons present as air pollutants, 0.2-8 ppm ethylene (in soil air) has been detected in some waterlogged soils in the field [198]. It is established that 1 ppm of this gas severely inhibits root extension of cereals, tomato, and tobacco [199]. Subsequent work has shown that a biological mechanism is mainly responsible for such release of ethylene under waterlogged conditions, with appreciable evolution only when the oxygen concentration falls below 1% [200]. Autoclaving soil decreased the production of ethylene by 90%, but the fact that exposure to 5 Mrad inhibited the yield by 50% was attributed to greater residual enzyme activity after irradiation [200].

In contrast, Rovira and Vendrell [201] found that irradiation of red-brown earth increased the evolution of ethylene: soil was stored for 10 weeks in waterlogged condition before receiving 6.2 Mrad, and at 2 hr after irradiation it formed twice as much ethylene as a nonsterile sample. The gas was still evolved at 10 weeks after γ treatment. Irradiation also stimulated the release of ethane, propane, propene, isobutane and butene, but in amounts that were not considered phytotoxic [201]. Ethylene can be produced by γ degradation of polyethylene and rubber; therefore, irradiation of soil in containers incorporating these materials should be avoided [200, 201]; it has been suggested that radiolytic attack on soil organic matter may be a contributory factor to formation of this gas [201].

Explanation of both inhibitory [200] and stimulatory [201] effects of radiation on the release of ethylene from soil must await further research. However, prolonged storage of soil at field capacity, as featured in the work of Rovira and Vendrell [201], is liable to depress the subsequent evolution of ethylene without the introduction of radiation [200]. Furthermore, this effect can be reversed by a period of air-drying before rewetting [200]. If

previously waterlogged and stored soil is used, one possibility is
that radiation acts in the same way as air-drying [156], raising
the yield of ethylene in relation to the control sample by improving
enzyme-substrate interaction upon damage to microorganisms. If no
initial stress to the formation of ethylene is imposed by soil storage conditions, the harmful effect of radiation is evident [200].
A broad comparison could thus be made with the stimulatory and inhibitory effects of radiation on nitrification under conditions of
restricted and ample ammonium substrate, respectively [32], as described in Section VI, B.

VII. CONCLUDING REMARKS

Treatment of soil with γ radiation or electron beams to achieve
partial or complete sterilization or to selectively eliminate specific organisms is established as a valuable technique applicable
to many aspects of soil research. Knowledge so far gained on the
microbiology and biochemistry of irradiated soils now permits more
reliable experimental procedures and, in particular, shows some
differential effects on soil nitrogen metabolism that may be utilized
to advantage. For example, biological denitrification of nitrate
to nitrite with further reduction and volatilization of products,
such as N_2O, is of economic importance to agriculture [152], and the
exposure of certain soils to 0.75 Mrad radiation appears to be a
useful pretreatment to study the biological reduction of nitrate
and nitrite to oxides of nitrogen and gaseous nitrogen at medium
soil moisture.

Nevertheless, our understanding of the way in which radiation
disturbs metabolic cycles in the soil ecosystem and upsets the
balance between groups of microorganisms is far from complete. The
results of prolonged exposure of soil to radiation in γ fields has
received relatively little attention compared with investigations
on the effects of acute treatments, and sensitive techniques that
are now available to record the release of gases, such as CO_2 and

oxides of nitrogen, from soil in the field [202, 203] could be applied.

Measurement of the activities of soil organisms by biochemical means [204] should help to resolve the effects of radiation on soil sulfur and manganese metabolism. Other worthwhile approaches to soil irradiation research and application have been mentioned in the course of this review.

REFERENCES

1. P. A. Cawse, *UKAEA Harwell Report R6061*, HMSO, London (1969).
2. H. P. Muller, *Atompraxis*, *14*, 480 (1968).
3. A. D. McLaren, *Soil Biol. Biochem.*, *1*, 63 (1969).
4. *Report ICRU*, *Handbook 62*, Natl. Bur. Standards, Washington,
    D.C., 1956.
5. S. Jefferson, in *Food Irradiation*, IAEA, Vienna, 1966.
6. F. Rogers and B. Whittaker, *UKAEA Harwell Report R6377*, HMSO,
    London (1970).
7. A. D. McLaren, R. A. Luse, and J. J. Skujins, *Soil Sci. Soc.
    Amer. Proc.*, *26*, 371 (1962).
8. S. Jefferson and C. G. Crawford, in *Radiosterilisation of
    Medical Products*, IAEA, Vienna, 1967.
9. E. E. Fowler, K. G. Shea, and G. R. Dietz, in *Food Irradiation*,
    IAEA, Vienna, 1966.
10. S. Jefferson and F. Rogers, in *Massive Radiation Techniques*
    (S. Jefferson, Ed.), Newnes, London, 1964.
11. D. de Zeeuw, *Euratom Rep.*, 4080e (1967).
12. The Patent Office, 25, Southampton Buildings, London WC2:
    patent 1133866, "Disinfection of Soils."
13. B. Whittaker, *UKAEA Harwell Report R3360*, HMSO, London (1964).
14. R. W. Clarke, in *Massive Radiation Techniques* (S. Jefferson,
    Ed.), Newnes, London, 1964.
15. J. C. Corey, D. R. Nielson, J. C. Picken Jr., and D. Kirkham,
    Environ. Sci. Tech., *1*, 144 (1967).
16. B. W. Clark and D. C. Coleman, *Pedobiologia*, *10*, 199 (1970).
17. G. R. Vela, Ph.D. Thesis, University of Texas, Austin, 1964.
18. G. R. Vela and O. Wyss, *J. Bacteriol.*, *89*, 1280 (1965).
19. S. Davison, S. A. Goldblith, and B. E. Proctor, *Nucleonics*,
    14, 34 (1956).
20. S. C. Sigaloff, *USAF School of Aviation Medicine Rep. 57-86*,
    Randolph AFB, Texas, 1957.
21. A. D. McLaren, L. Reshetko, and W. Huber, *Soil Sci.*, *83*, 497
    (1957).
22. M. J. Stocklasa, *C. R. Acad. Sci.*, *155*, 1096 (1912).
23. M. J. Stocklasa, *C. R. Acad. Sci.*, *156*, 153 (1913).
24. M. J. Stocklasa, *C. R. Acad. Sci.*, *157*, 879 (1913).
25. E. Kayser, *C. R. Acad. Sci.*, *172*, 1133 (1921).
26. E. Kayser and H. Delaval, *C. R. Acad. Sci.*, *179*, 110 (1924).
27. R. M. Whelden, E. V. Enzmann, and C. P. Haskins, *J. Gen.
    Physiol.*, *24*, 789 (1941).
28. E. N. Sokurova, *Izv. Akad. Nauk SSSR.*, *Ser. Biol.*, *6*, 35 (1956).
29. G. H. Peterson, *Soil Sci.*, *94*, 71 (1962).
30. P. A. Cawse and K. M. Mableson, *Comm. Soil Sci. Plant Res.*, *2*,
    421 (1972).
31. P. A. Cawse and A. H. Cornfield, *Soil Biol. Biochem.*, *3*, 111
    (1971).
32. P. A. Cawse, *J. Sci. Food Agri.*, *19*, 395 (1968).

33. R. J. Davis, V. L. Sheldon, and S. I. Auerbach, *J. Bacteriol.*, *72*, 505 (1956).
34. S. I. Auerbach, *Ecology*, *39*, 522 (1958).
35. C. A. Edwards, in *Proc. 2nd Natl. Symp. Radioecol.*, Ann Arbor, Michigan, 1967.
36. F. J. Ley and A. Tallentire, *Pharmacol. J.*, *193*, 59 (1964).
37. D. E. Lea, in *Actions of Radiations on Living Cells*, 2nd Ed., Univ. Press, Cambridge, 1955.
38. I. L. Stevenson, *Plant Soil*, *8*, 170 (1956).
39. D. A. Soulides and F. E. Allison, *Soil Sci.*, *91*, 291 (1961).
40. F. J. Ley and A. Tallentire, *Pharmacol. J.*, *195*, 216 (1965).
41. C. F. Eno and H. Popenoe, *Soil Sci. Soc. Amer. Proc.*, *28*, 533 (1964).
42. J. P. Voets, M. Dedeken, and E. Bessems, *Naturwissenschaften*, *16*, 476 (1965).
43. J. Boyer, A. Combeau, A. Grauby, and A. Thomas, in *Bulletin D'Informations Scientifiques et Techniques*, no. 106, CEA, Paris, 1966.
44. G. G. Jayson, D. C. Pickering, W. Markham, and M. S. Parker, *Naturwissenschaften*, *8*, 185 (1965).
45. N. E. Jackson, J. C. Corey, L. R. Frederick, and J. C. Picken Jr., *Soil Sci. Soc. Amer. Proc.*, *31*, 491 (1967).
46. G. E. Stapleton, *Bacteriol. Rev.*, *19*, 26 (1955).
47. H. F. Birch, *Plant Soil*, *10*, 9 (1958).
48. A. P. Casarett, in *Radiation Biology* (C. P. Swanson, Ed.), Prentice-Hall, Englewood Cliffs, New Jersey, 1966.
49. J. A. Ghormley, *Rad. Res.*, *5*, 247 (1956).
50. G. Scholes and J. Weiss, *Nature (London)*, *185*, 305 (1960).
51. S. Okada, in *Organic Peroxides in Radiobiology*, Pergamon, London, 1958.
52. G. B. Kashkina and Yu. D. Abaturov, *Radiobiologiya*, *7*, 263 (1967).
53. M. D. Socolofsky and O. Wyss, *J. Bacteriol.*, *84*, 119 (1962).
54. M. Monib and M. N. Zayed, *J. Appl. Bacteriol.*, *26*, 35 (1963).
55. V. Vinter, *Nature (London)*, *183*, 998 (1959).
56. V. Vinter, *Nature (London)*, *189*, 589 (1961).
57. H. Dertinger and H. Jung, in *Molecular Radiation Biology*, Longman, London, 1970.
58. B. A. Bridges and T. Horne, *J. Appl. Bacteriol.*, *22*, 96 (1959).
59. G. Stehlik and K. Kaindl, in *Food Irradiation*, IAEA, Vienna, 1966.
60. D. B. Menzel, *Ann. Rev. Pharmacol.*, *10*, 379 (1970).
61. Z. I. Kertesz and G. F. Parsons, *Science*, *142*, 1289 (1963).
62. H. P. Muller and L. Schmidt, *Arch. Microbiol.*, *54*, 70 (1966).
63. A. M. Kuzin, in *Radiation Biochemistry* (M. R. Quastel, Ed.), Israel Programme for Scientific Translations, Jerusalem, 1964.
64. T. Ojima, H. Toratani, and H. Fujimoto, *Ann. Rept. Rad. Cent. Osaka*, *4*, 107 (1963).
65. H. J. M. Bowen and P. A. Cawse, *Soil Sci.*, *97*, 252 (1964).

66. G. Stotzky and J. L. Mortensen, *Soil Sci. Soc. Amer. Proc.*, *23*, 125 (1959).
67. M. R. Roberge and R. Knowles, *Naturaliste Can.*, *94*, 221 (1967).
68. J. P. Skou, *J. Gen. Microbiol.*, *28*, 521 (1962).
69. L. F. Johnson and T. S. Osborne, *Can. J. Bot.*, *42*, 105 (1964).
70. L. S. Lur'e and L. G. Ter-Simonyan, *Lenin Acad. Agri. Sci. Proc.*, *24*, 28 (1959).
71. B. Mosse, D. S. Hayman, and G. J. Ide, *Nature (London)*, *224*, 1031 (1969).
72. B. Mosse and D. A. Hayman, *New Phytol.*, *70*, 29 (1971).
73. O. R. Eylar, Jr., and E. L. Schmidt, *J. Gen. Microbiol.*, *20*, 473 (1959).
74. M. Ishaque, A. H. Cornfield, and P. A. Cawse, *Plant Soil*, *35*, 201 (1971).
75. P. A. Cawse and T. White, *J. Agri. Sci.*, *72*, 331 (1969).
76. G. B. Kashkina and Yu. D. Abaturov, *Radiobiologiya*, *9*, 460 (1969).
77. S. E. Gochenaur, *AED Conf. 298-003*, Bloomington, Indiana, 1970.
78. M. Witkamp, in *A Tropical Rain Forest* (H. D. Odum, Ed.), USAEC TID-24270, Oak Ridge, Tennessee, 1970.
79. J. R. Holler and G. T. Cowley, in *A Tropical Rain Forest* (H. D. Odum, Ed.), USAEC TID-24270, Oak Ridge, Tennessee, 1970.
80. L. M. Shields, L. W. Durrell, and A. H. Sparrow, *Ecology*, *42*, 440 (1961).
81. L. W. Durrell and L. M. Shields, *Mycologia*, *52*, 636 (1960).
82. D. F. Splitstoesser and L. M. Massey, Jr., *Appl. Microbiol.*, *15*, 646 (1967).
83. J. J. Perry, in *A Tropical Rain Forest* (H. D. Odum, Ed.), USAEC TID-24270, Oak Ridge, Tennessee, 1970.
84. V. V. Bernard and I. T. Geller, *Agrobiologiya*, No. 4, 610 (1962).
85. S. N. Blumenfeld, *USAEC Report RM-4827-TAB* (1966).
86. R. Stanovick, J. Giddens and R. A. McCreery, *Soil Sci.*, *92*, 183 (1961).
87. H. J. M. Bowen and P. A. Cawse, in *Radioisotopes in Soil-Plant Nutrition Studies*, IAEA, Vienna, 1962.
88. G. Skyring and J. P. Thompson, *Plant Soil*, *24*, 289 (1966).
89. A. D. Rovira and G. D. Bowen, *Austral. J. Soil Res.*, *7*, 57 (1969).
90. R. T. Wood and L. E. Casida, *Bacteriol. Proc.*, *67*, 2 (1967).
91. H. Lees and J. H. Quastel, *Biochem. J.*, *40*, 815 (1946).
92. G. R. Vela, *Texas J. Sci.*, *20*, 315 (1969).
93. P. A. Cawse and D.V. Crawford, *Nature (London)*, *216*, 1142 (1967).
94. A. Rambelli, *Pub. Cent. Sper. Agr. For.*, *6*, 117 (1962).
95. A. Rambelli, *Pub. Cent. Sper. Agr. For.*, *7*, 127 (1963).
96. T. Titani, M. Kondo, H. Suguro, and T. Takahashi, in *Proc. 2nd Int. Conf. on Peaceful Uses of Atomic Energy*, UN, New York, 1958.
97. F. J. Ley, *Int. J. Appl. Radiat. Isot.*, *14*, 38 (1963).
98. H. Popenoe and C. F. Eno, *Soil Sci. Soc. Amer. Proc.*, *26*, 164 (1962).
99. M. B. E. Godward, *Nature (London)*, *185*, 706 (1960).

100. J. J. Skujins, L. Braal, and A. D. McLaren, *Enzymologia*, *25*, 125 (1962).
101. J. R. Ramirez-Martinez and A. D. McLaren, *Enzymologia*, *31*, 23 (1966).
102. M. R. Roberge and R. Knowles, *Can. J. Soil Sci.*, *48*, 355 (1968).
103. J. J. Skujins and A. D. McLaren, *Soil Biol. Biochem.*, *1*, 89 (1969).
104. A. D. McLaren, *Rec. Progr. Microbiol.*, *8*, 221 (1963).
105. G. R. Vela and O. Wyss, *Bacteriol. Proc.*, *62*, 24 (1962).
106. M. R. Roberge, *Can. J. Microbiol.*, *16*, 865 (1970).
107. J. J. Skujins, in *Soil Biochemistry* (A. D. McLaren and G. H. Peterson, Eds.), Vol. 1, Dekker, New York, 1967.
108. M. R. Roberge, *Can. J. Microbiol.*, *14*, 999 (1968).
109. P. E. Le R. Van Niekerk, *South African J. Agri. Sci.*, *7*, 131 (1964).
110. D. S. Jenkinson and D. S. Powlson, *Soil Biol. Biochem.*, *2*, 99 (1970).
111. D. A. Barber, in *Nitrogen-15 in Soil-Plant Studies*, IAEA, Vienna, 1971.
112. D. A. Barber and G. J. Benians, in *ARC Letcombe Laboratory Ann. Report (1968)*, ARCRL 19, Wantage, England, 1969.
113. A. D. McLaren and J. J. Skujins, *Can. J. Microbiol.*, *9*, 729 (1963).
114. P. A. Cawse, *UKAEA Harwell Report PR/HPM 15* HMSO, London (1971).
115. P. A. Cawse, *UKAEA Harwell Report PR/HPM 16* HMSO, London (1972).
116. G. H. Peterson, *Can. J. Microbiol.*, *8*, 519 (1962).
117. P. O. Salonius, J. B. Robinson, and F. E. Chase, *Plant Soil*, *27*, 239 (1967).
118. R. J. Hervey and Z. M. Williams, *Texas Rept. Biol. Med.*, *21*, 450 (1963).
119. C. Bloomfield, *J. Soil Sci.*, *23*, 1 (1972).
120. D. C. Coleman and A. MacFadyen, *Oikos*, *17*, 62 (1966).
121. D. S. Jenkinson, T. Z. Nowakowski, and J. D. D. Mitchell, *Plant Soil*, *36*, 149 (1972).
122. A. H. Merzari and H. Broeshart, in *Isotope Studies on the Nitrogen Chain*, IAEA, Vienna, 1968.
123. R. Y. Stanier, *Soil Sci.*, *53*, 479 (1942).
124. J. Meiklejohn, *Plant Soil*, *3*, 88 (1951).
125. R. J. Berry, P. R. Hills, and W. Trillwood, *Int. J. Rad. Biol.*, *9*, 559 (1965).
126. I. Gottlieb and P. Markakis, *Rad. Res.*, *36*, 55 (1968).
127. J. Schubert, J. A. Watson, and E. R. White, *Int. J. Rad. Biol.*, *13*, 485 (1967).
128. U. C. Gupta, in *Soil Biochemistry* (A. D. McLaren and G. H. Peterson, Eds.), Vol. 1, Dekker, New York, 1967.
129. E. J. Gibson, *Report RR 1859* (1959), Warren Spring Laboratory, Stevenage, England.
130. D. M. Smith and R. Y. Mixer, *Rad. Res.*, *11*, 776 (1959).
131. H. J. M. Bowen and P. A. Cawse, *Soil Sci.*, *98*, 358 (1964).
132. P. Bovard and A. Grauby, *Energie Nucléaire*, *5*, 149 (1963).

133. P. A. Cawse, *J. Sci. Food Agri.*, *18*, 388 (1967).
134. C. F. Eno and H. Popenoe, *Soil Sci. Soc. Amer. Proc.*, *27*, 299 (1963).
135. M. W. Loutit, *Soil Biol. Biochem.*, *2*, 131 (1970).
136. P. A. Cawse, *UKAEA Harwell Report PR/HPM 14*, HMSO, London (1970).
137. M. Alexander, in *Introduction to Soil Microbiology*, Wiley, New York, 1967.
138. B. R. Singh and Y. Kanehiro, *J. Sci. Food Agri.*, *21*, 61 (1970).
139. J. J. Skujins and A. D. McLaren, *Science*, *158*, 1569 (1967).
140. T. White, HND Thesis, Trent Polytechnic, Nottingham, England, 1969.
141. D. S. Jenkinson, *J. Soil Sci.*, *17*, 280 (1966).
142. T. F. Hering, *Trans. Brit. Mycol. Soc.*, *50*, 267 (1967).
143. C. F. Burgos, *Ceiba*, *10*, 53 (1964).
144. P. A. Cawse and A. H. Cornfield, *Soil Biol. Biochem.*, *4*, 497 (1972).
145. P. A. Cawse and A. H. Cornfield, *Soil Biol. Biochem.*, *1*, 267 (1969).
146. C. L. Valera, *Diss. Abstr.*, *20*, 1929 (1959).
147. C. L. Valera and M. Alexander, *Plant Soil*, *15*, 268 (1961).
148. P. A. Cawse and T. White, *J. Agri. Sci.*, *73*, 113 (1969).
149. F. E. Allison, *Soil Sci.*, *96*, 404 (1963).
150. H. Nommik, *Acta Agri. Scand.*, *6*, 195 (1956).
151. N. E. R. Campbell and H. Lees, in *Soil Biochemistry* (A. D. McLaren and G. H. Peterson, Eds.), Vol. 1, Dekker, New York, 1967.
152. R. D. Hauck, *Trans. 9th Intern. Congr. Soil Sci.*, *2*, 475 (1968).
153. D. M. Ekpete and A. H. Cornfield, *Nature (London)*, *201*, 322 (1964).
154. F. E. Broadbent, *Soil Sci.*, *72*, 129 (1951).
155. F. E. Broadbent and B. F. Stojanovic, *Soil Sci. Soc. Amer. Proc.*, *16*, 359 (1952).
156. P. A. Cawse and D. S. Sheldon, *J. Agri. Sci.*, *78*, 405 (1972).
157. B. R. Singh, A. S. Agarwal, and Y. Kanehiro, *Soil Sci.*, *108*, 85 (1969).
158. H. A. Mahlman and G. K. Schweitzer, *J. Inorg. Nucl. Chem.*, *5*, 213 (1958).
159. M. Daniels and E. E. Wigg, *J. Phys. Chem.*, *71*, 1024 (1967).
160. A. K. Pikaev, in *Pulse Radiolysis of Water and Aqueous Solutions*, Indiana Univ. Press, Bloomington, 1967.
161. M. Daniels, *J. Phys. Chem.*, *73*, 3710 (1969).
162. J. M. Bremner and D. W. Nelson, *Trans. 9th Int. Congr. Soil Sci.*, *2*, 495 (1968).
163. E. Grossbard, *Mededel. Fac. Landbouw. Wetenschappen Gent*, *35*, 515 (1970).
164. M. N. Meisel, *Meetings of Biological Sciences Division, Peaceful Uses of Atomic Energy, Moscow*, *106* (1955).
165. M. R. Roberge, *Soil Sci.*, *111*, 124 (1971).
166. R. H. Johnsen, *J. Phys. Chem.*, *63*, 2041 (1959).
167. M. Witkamp, USAEC Report ORNL-TM-29 (1961).

168. C. A. I. Goring and F. E. Clark, *Soil Sci. Soc. Amer. Proc.*, *16*, 7 (1952).
169. H. van Groenewoud, *Proc. Can. Phytopath. Soc.*, No. 26, 12 (1958).
170. L. W. Getzin and I. Rosefield, *Agri. Food Chem.*, *16*, 598 (1968).
171. P. C. Lippold, J. S. Cleere, L. M. Massey Jr., J. B. Bourke, and A. W. Avens, *J. Econ. Entomol.*, *62*, 1509 (1969).
172. K. M. Bowmer, Ph.D. thesis, University of Nottingham, England, 1969.
173. I. J. Graham-Bryce, *J.Sci.Food Agri.*, *18*, 72 (1967).
174. H. D. Skipper, C. M. Gilmour, and W. R. Furtick, *Soil Sci. Soc. Amer. Proc.*, *31*, 653 (1967).
175. D. E. Armstrong, G. Chesters, and R. F. Harris, *Soil Sci. Soc. Amer. Proc.*, *31*, 61 (1967).
176. D. G. Cummins and R. A. McCreery, *Soil Sci. Soc. Amer. Proc.*, *28*, 390 (1964).
177. J. H. Warcup, *Soils Fert.*,*20*, 1 (1957).
178. P. A. Cawse, private communication.
179. R. G. Florence, Ph.D. Thesis, University of Sydney, Australia, 1961.
180. J. M. Ulrich, R. A. Luse, and A. D. McLaren, *Physiol. Plant.*, *17*, 683 (1964).
181. M. W. Loutit, R. S. Malthus and J. S. Loutit, *New Zealand J. Agri. Res.*, *11*, 420 (1968).
182. M. W. Loutit, *Trans. 9th Int. Congr. Soil Sci.*, *3*, 491 (1968).
183. P. A. Cawse and J. H. Radley, *Rad. Bot.*, *10*, 491 (1970).
184. K. L. Webb and R. H. Hodgson, *Science*, *132*, 1762 (1960).
185. G. D. Bowen and A. D. Rovira, *Nature (London)*, *191*, 936 (1961).
186. A. D. Rovira and G. D. Bowen, *Plant Soil*, *25*, 129 (1966).
187. G. W. Winsor, *Rept. Progr. Appl. Chem.*, *49*, 326 (1964).
188. F. T. Bingham, H. D. Chapman, and A. L. Pugh, *Soil Sci. Soc. Amer. Proc.*, *18*, 305 (1954).
189. D. S. Curtis, *Soil Sci.*, *68*, 441 (1949).
190. F. V. Widdowson and A. Penny, in *Rep. Rothamsted Exp. Stn. 1969*, Pt. 2, 1970.
191. E. Griffiths and R. G. Burns, *Plant Soil*, *28*, 169 (1968).
192. D. S. Sheldon, HND Thesis, Trent Polytechnic, Nottingham, England, 1972.
193. C. C. Delwiche and J. Wijler, *Plant Soil*, *7*, 113 (1956).
194. D. J. Greenland, *J. Agri. Sci.*, *58*, 227 (1962).
195. H. L. Jensen, *Trans. 4th Int. Cong. Soil Sci.*, *1*, 165 (1950).
196. W. A. Rice, E. A. Paul, and L. R. Wetter, *Can. J. Microbiol.*, *13*, 829 (1967).
197. F. B. Abeles, L. E. Craker, L. E. Forrence, and G. R. Leather, *Science*, *173*, 914 (1971).
198. K. A. Smith and R. S. Russell, *Nature (London)*, *222*, 769 (1969).
199. K. A. Smith and P. D. Robertson, *Nature (London)*, *234*, 148 (1971).
200. K. A. Smith and S. W. F. Restall, *J. Soil Sci.*, *22*, 430 (1971).

201.   A. D. Rovira and M. Vendrell, *Soil Biol. Biochem.*, *4*, 63
        (1972).
202.   J. W. McGarity and R. D. Hauck, *Soil Sci.*, *108*, 335 (1969).
203.   J. R. Burford and R. J. Millington, *Trans. 9th Int. Cong.
        Soil Sci.*, *2*, 505 (1968).
204.   A. D. McLaren, in *Soil Biochemistry* (A. D. McLaren and G. H.
        Peterson, Eds.), Vol. 1, Dekker, New York, 1967.

CHAPTER 6

MICROBIAL GROWTH AND CARBON TURNOVER

George H. Wagner
Department of Agronomy
University of Missouri
Columbia, Missouri

I. INTRODUCTION

The soil functions as a natural waste disposal system for debris from life forms growing on and within it. Organic materials

of various origins and with widely differing compositions and structures temporarily accumulate in the soil. They reside here for dur-
ations ranging from a few days to thousands of years and come under
attack by a population of heterotrophic soil microorganisms.

Clark [1] recently reviewed growth of bacteria in soil and
emphasized the specialization that microorganisms have developed
for utilizing energy-yielding materials. A cosmopolitan substrate
is presented to the microbes and the utilization of it is governed
by the ecological conditions that the substrate encounters. In a
review dealing with the growth of fungi in soil, Waid [2] noted
that fungal growth is selectively related to the substrate available.
Fungi generally account for a large amount of microbial cell material in soil but little is known about the duration of this or other
microbial tissue in soil.

The longevity of active vegetative cells of microorganisms in
soil and the relationship of this longevity to carbon turnover are
considered here. Research on this subject is generally recent and
rather meager. A complete understanding of the subject must await
subsequent years of imaginative effort. Some attention is given
to recent kinetic studies dealing with the efficiency of growth
of microorganisms in aquatic environments and in sewage sludge,
where the concepts may be relevant to microbial growth in soil.

The occurrence and duration of carbon in microbial tissues is
a fundamental part of the larger concept of carbon in the biosphere
and the carbon cycle in nature. A general view of the carbon cycle
is presented to provide perspective for quantitative considerations
of microbial growth and carbon turnover.

II. THE CARBON CYCLE

The element carbon makes up the framework of biological systems. Biosynthetic processes link carbon to itself and to other
elements in the assembly of a vast collection of organic compounds

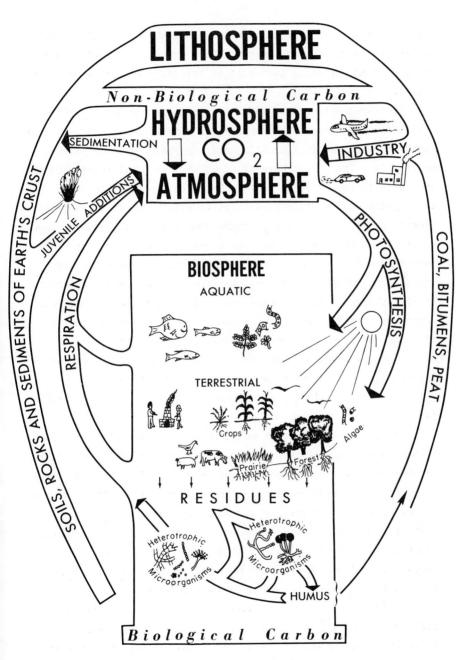

FIG. 1.   The carbon cycle in nature.

essential for life. Carbon enters the biosphere through photosynthesis and leaves through respiration, completing a cycle. A schematic diagram depicting the cycle of carbon in nature is given in
Fig. 1.

Photosynthetic plants and a few chemoautotrophic organisms
assimilate CO_2 from the atmosphere and convert it into organic compounds of high energy content. Heterotrophic CO_2 utilization is
negligible. The annual harvest of CO_2 by the total vegetation of
the earth approximates 3.5×10^{13} kg of carbon [3]. About one-half
or slightly more of this production is attributed to terrestrial
plants [3, 4]. There is disagreement on this point, however, and
estimates of much higher production for marine environments have
been reported [5]. The carbon in photosynthetic products has a
transitory journey through the biosphere. Some carbon may enter
several phases of life successively. The energy in the original
compounds is passed along to various forms of heterotrophic life
that convert it by biochemical processes, and in the course of
these, CO_2 is respired under aerobic conditions. Total carbon in
living organisms is estimated to be between 3 and 5×10^{14} kg [3, 4].
Most of this biologically fixed carbon occurs in forest vegetation
[4].

The release of CO_2 back to the atmosphere results from plant,
animal, and microbial respiration. The major respiratory activity
is attributed to heterotrophic microorganisms [3]. Members of the
microbial population possess differing enzyme capabilities matching
the diversity of the organic compounds occurring in the biosphere.
Initially, decay involves attack by the microbes upon the residues
of higher plants and animals. In turn the soil microbes feed on
themselves and upon humic components as the last of the energy that
entered the biosphere by photosyntehsis is reaped.

Total return of CO_2 by decomposition and respiration is estimated at 3.5×10^{13} kg of carbon per year [6]. This estimate
indicates that the quantity returned is in balance with that fixed
by photosynthesis. Other estimates [4], which distinguish the

terrestrial from the aquatic arms of the cycle, are not in complete
agreement with this quantity but support the suggested balance between fixation and release of CO_2. Both net annual photosynthetic
fixation and respiratory liberation of CO_2 equal nearly 5% of the
amount present in the atmospheric reservoir [6]. Aquatic photosynthesis and respiration, however, function essentially on dissolved gasses as a self-contained cycle in the sea [4]. Nevertheless, the cycles are linked because CO_2 in the atmosphere is in
equilibrium with that in the sea [7]. The hydrosphere contains
somewhere between 50 and 60 times the amount of CO_2 found in the
atmosphere [3, 4].

Release of CO_2 by the burning of fossil fuels has enriched the
atmosphere within the last century [4] because this avenue of the
cycle has no apparent counterbalance. A buffering effect of the
oceans exists, however, and it is estimated that about two-thirds
of the CO_2 released from fossil fuels has gone into the oceans.
Natural, slow processes that account for circulation of carbon in
its various geological and oceanic forms include the transfer of
carbon to the lithosphere by precipitation or loss to sediments and
the liberation of juvenile carbon dioxide to the atmosphere by
volcanoes and mineral springs [4, 8].

Humus is a special carbonaceous product of the decay process
in soil that possesses unique durability, even though occurring in
an environment seething with microorganisms. Humic materials include a spectrum of highly altered remnants of plant, animal, and
microbial tissues along with products synthesized by the attacking
microorganisms, all occurring collectively as an amorphous, dark
colored, more or less stable fraction of the organic matter in
soil [9]. Although humus is relatively resistant to biologic soil
attack, it is eventually degraded to liberate CO_2, probably by the
autochthonous soil microorganisms. Under special conditions that
limit decomposition, some carbon from humus or plant residues may
possibly, as in past geologic history [4], be converted to bitumens
and coal. Decomposition of vegetation under fresh bodies of water

results in the accumulation of peat. Relative to the total amount
of carbon in the lithosphere, this present rate of transfer is inconsequential. However, the existence of a large reservoir of
fossil fuel, estimated at 50 times the amount of carbon in all
living organisms [4], suggests that at some time a significant accumulation occurred when net photosynthesis exceeded respiration.

In addition to a global view of the cycling of carbon in nature, a further valuable perspective would be that of carbon flow
in specific ecosystems. The fragmentary quantitative information
of this type that is presently available will be used, where relevant, in the development of the carbon turnover theme of this
chapter.

It is hoped that a well-defined assessment of the flow of carbon for a total system will become available in the near future.
The staff connected with the Grassland Biome of the United States
International Biological Program is engaged in this kind of activity
[10]. An objective of this research team is to develop a total-
system model of the biomass dynamics for a grassland that, via parameter change, can represent the sites in the Grassland Biome Network. Using the model designated ELM 1.0 at its present stage of
development, annual dry matter flow for the Pawnee Grassland in
Colorado has been obtained and is summarized as a compartment model
in Fig. 2. The output simulates 1 year, starting January 1. The
driving variables were generated stochastically and are typical of
the Pawnee site with annual rainfall totaling 13.2 in. Initial and
final values are presented for each compartment and are estimated
from field measurements. Net flows of dry matter for the year, presented along the arrows, are taken from graphical output and are
not as exact. This accounts for the apparent failure of the flows
into and out of a compartment to exactly equal the change in dry
matter content of the compartment. One deficiency of the model is
the omission of a compartment for humus. The authors are aware of
this problem but have chosen to show flow from the compartment for
microbes directly to source and sink because no suitable experimental

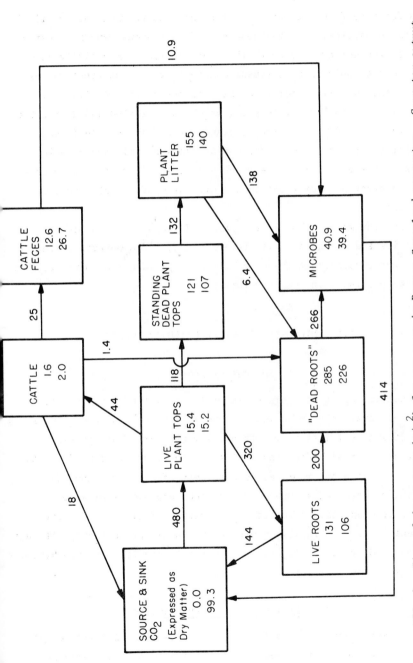

FIG. 2. Flow of dry matter (g/m^2) for one year in Pawnee Grassland ecosystem. Computer output from Grassland Biome Model ELM 1.0 summarized according to compartments with the top value in each compartment as the initial dry matter level, January 1; the bottom value as the final level 1 year later; and the values along the arrows showing dry matter flow between compartments [10].

data defining flow into and out of a humus compartment are available
for the ecosystem. It has been assumed that humus remains constant
during time periods for which the model applies. Especially relevant to the topic under discussion here is the compartment for microbes and the dry matter flow into and out of it. A microbial
biomass of approximately 40 g dry wt/m^2 is indicated and during the
year this is fed by dead roots, plant litter, and cattle feces,
totalling 415 g/m^2. The model output shows a small decrease in
biomass at the end of the year and a release of CO_2 through microbial respiration equivalent to the dry matter flow into the compartment.

Similar modeling activities are underway in various other lab-
oratories and these deal with forest, tundra, and desert ecosystems.
It is anticipated that these efforts will yield useful material that
will be published in the near future.

III. MICROBIAL GROWTH RELATED TO ENERGY

A. Sources of Energy

Heterotrophic microorganisms of the soil find a variety of
materials available as sources of energy. Plant residues, consisting
of root and leaf litter from trees, shrubs, herbs, and grasses,
along with animal bodies and wastes are continually being fed into
soil. Living plant roots exude a variety of organic compounds into
the rhizosphere and sloughed root tissues are also available. Appreciable quantities of microbial products and tissues are ever
present in soil. Finally, humic materials are present and, although
not as readily available, they may serve to nourish the autochthonous
members of the soil population.

B. Distribution of Energy Sources

Organic materials available as energy sources for microorganisms
are heterogenously dispersed in soil. This is most universally

recognized in the rhizosphere, where high populations are largely attributed to a high concentration of readily available substrate in the region of soil at a root surface [11, 12]. Both sloughed root debris and exuded organic materials serve as energy sources for rhizosphere microorganisms. Clark and Paul [13] extrapolated the measurements of cell debris from pea and oat plants to arrive at a value for the growing season of 4 g of sloughed material per square meter. They concluded from their examination of the literature that the amount of soluble exudate exceeds particulate debris and ranges upward from 1% of the organic matter in the mature plant, with an average of nearly 10%.

Beyond the rhizosphere influence, microbial cell distribution is also focal. Russian workers have shown a nondiffuse distribution of bacteria in soil for various sample sizes that ranged from the field scale downward to tiny aggregates that contained bacteria in microfoci [14]. By direct microscopic observations in a sand dune soil, Gray et al. [15] showed that the bacterial count was considerably higher on organic particles than on sand particles, which were observed to be only sparsely colonized. About 60% of the bacteria occurred on the organic particles, even though these particles accounted for only 15% of the colonizable particle surface in the soil. A further report [16] stated that colonies occurring on organic particles were larger and contained more cells per colony than those on sand particles. It was estimated from these observations that 0.04% of the total particle surfaces were covered by bacteria.

C. Pattern of Growth

It has been suggested [17] that sites of growth in the soil may be reflected by growth patterns associated with different members of the population as they invade and colonize substrates at their disposal. Gray and Williams [18] have recently described these patterns of growth, implying a relationship to substrate distribution and observing that growth is common on substrate surfaces

or may penetrate within substrates. They note that nonmigratory
unicellular microbes exist on surfaces that provide substrate and
that migratory unicellular organisms do not show active movement
through soil but along continuous surfaces, such as roots. A
restricted hyphal pattern of growth in soil, according to their
scheme, is limited to sites on and within particulate substrate
and locally spreading hyphal growth in soil shows site colonization
spreading from a particulate substrate, whereas mycelial strands
that grow through soil are considered to be moving from one substrate
site to another. A diffuse spreading hyphal pattern of growth un-
associated with particulate substrate, they assume, is using nutrients in the soil solution.

Observations of microorganisms in soil only reveal existing
structures and it is difficult to determine whether or not active
microbial growth is occurring. Microbial cells seen by direct microscopy may be dormant or persistent dead structures that merely
indicate past growth.

Microbial growth rates in nature are the subject of a recent
review by Brock [19]. He discusses a variety of methods for examining microbial growth in various natural habitats. Attention is
called to the use of pedoscopes [20] and the use of incident illum-
ination in direct microscopy of undisturbed soil particles [21].
Brock [19] challenges microbial ecologists to consider methods that
involve labeling of cell DNA with radioactive thymidine, heretofore
used in studying mammalian systems, and to develop the necessary
modifications so these methods may be used to study microorganisms
in natural habitats.

D. Microbial Efficiency and Growth

Soil respiration measurements in aerobic systems, in contrast
to microscopic observations of microorganisms, directly indicate
total activity. Although it is difficult to resolve microbial and
root respirations in soils in which there is standing vegetation,
both oxygen consumption and carbon dioxide production measurements

are useful indices of microbial activity [22]. Decay of plant and
animal tissues in aerobic soil involves transformations that yield
oxidized forms of carbon, the ultimate of which is CO_2. The energy
released by the oxidation drives the metabolic processes of the
microbes and some of the carbon in the substrate is utilized to synthesize new microbial tissues. A low efficiency of utilizing the
substrate carbon for new synthesis would coincide with rapid turnover and more carbon would be liberated as CO_2. Widely quoted estimates by Waksman [23] suggest that during decomposition by fungi,
30-50% of the carbon metabolized is used to form new mycelium, whereas aerobic bacteria assimilate 20-40% and anaerobic bacteria incor-
porate only 2-5% of the substrate carbon into new cells.

Lees and Porteous [24] studied decomposition of several simple
organic substrates. Their results indicate that the soil microflora assimilate from 25 to 33% of the substrate carbon. The proportion of carbon assimilated in maximal aerobic growth probably
exceeds that reported by Lees and Porteous. Terroine and Wurmser
[25] reported a production of 0.44 g of *Aspergillus niger* per gram
of sugar consumed. Assuming the mycelium contained 55% carbon, the
proportion of carbon assimilated would equal 60%. Siegel and
Clifton [26] reported the ratio of carbon assimilated to carbon
utilized for *Escherichia coli* to be 0.56-0.60. Payne [27] concluded, after evaluating recent research, that bacteria in oxidative
consumption of organic substrates between the initiation of growth
and the upper log phase assimilate about 60% of the carbon utilized.
Assimilation is lower for hydrocarbon-utilizing organisms, for
which no more than 40% of the carbon utilized is assimilated [28].
Blevins and Perry [29] have concluded that the efficiency of energy
transfer into cell material for a hydrocarbon-utilizing organism
is about one-half that of other organisms.

The assimilation of up to 70% of substrate carbon was shown
by Camp [30] to be theoretically possible. He calculated the heat

of formation of bacterial cell material from an empirical formula
and the heat of combustion of bacteria. An estimated requirement
of 0.95-1.95 kcal/g of dry cell material was obtained. Using an
average requirement of 1.5 kcal/g and assuming aerobiosis of sugar
he concluded that each gram of bacteria synthesized would require
1.42 g of glucose, 30% of which would be oxidized for energy and the
remainder used in synthesis of the bacterial cells.

In a review of energy yields and growth of heterotrophs, Payne
[27] has correlated maximum limits of growth with several functions
of available energy. He estimated yield of bacterial cells at 0.118
g dry wt/kcal of energy involved. The value was shown experimentally
to apply to the growth of pure cultures of both aerobic and anaerobic organisms and to aerobic mixed cultures derived from sewage and
from soil. A second relationship between yield and energy was defined in his approximation that cell yields of 3 g may be obtained
for each electron initially available in the substrate for transfer to oxygen or incorporation into cellular material. This yield
estimate is restricted to proliferating bacteria, which completely
consume the substrate and produce no significant products other than
cells and CO_2. A third relationship noted was that yield of cells
per mole of adenosine triphosphate (ATP) formed during growth
equalled approximately 10.5 g. The estimate held for anaerobic
growth in which ATP production was calculated from prior knowledge
of catabolic pathways, and for aerobic cultures, in which various
procedures for estimating ATP were employed. The estimates suggested
by Payne were in agreement with average yields calculated from many
different experiments, although it should be noted that considerable
variation existed among the individual values. A study of cell yield
for heterogeneous microbial populations of sewage origin grown on
glucose under well-defined experimental conditions led Gaudy and
Ramanathan [31] to conclude that an inherent variability prohibits
any precise definition of a yield coefficient.

IV. MICROBIAL GROWTH AND PLANT PRODUCTIVITY

Total annual organic debris available for support of heterotrophic soil microorganisms is roughly equal to annual plant production on earth. Total carbon in the forests of the earth is rather
constant from year to year, with the average life of a tree being
30 years [4]. A mature forest is considered by Brock [32] to be a
steady state ecosystem in which net production equals decomposition.
Burges [33] used annual litter fall plus annual death of roots as
an estimate of material being decomposed annually. For grasslands,
living plant biomass in the standing crop at the end of the growing
season may be taken as an approximation of the amount of material
being decomposed annually [13]. The formation of humus, which has
a long-term tenure, does not alter the annual balance sheet of
production and decomposition. On an annual basis humus formation
also is essentially equal to humus decomposition [34]; its equilibrium level varies among soils and is dependent upon climate and
vegetation [35].

A. Estimates of Plant Productivity

Annual net productivity on an energy basis for a tall-grass
prairie in Missouri has been estimated at 4.351×10^6 cal/m^2, representing 992 g of plant biomass [36]. Considering various vegetative covers, Gray [17] estimated that dry matter production aver-
aged 1 kg (4800 kcal)/m^2/year. A rather constant net primary production was postulated [37] because increased photosynthesis associated with a particular latitude was observed to be balanced by
increased respiration. Two forests differing considerably in gross
photosynthetic production were both estimated to have a net annual
production of 1.35 kg dry matter per square meter [17]. These estimates take into account mean annual increments of stems and
branches, the production of stem and branch litter, the increase
in butt weights, the production of leaf and other litter, and the

annual production of roots. Adjustments were made for the amount
of organic matter washed from the soil canopy and for material consumed by herbivores and removed from the site.

Below-ground production of tree and shrub roots has been estimated to be about 20% of the above-ground production [38]. For the
Missouri prairie, root biomass was 1449 g/m^2 prior to resumption of
growth in the spring and increased to 1901 g/m^2 at the end of the
growing season [39]. This annual increase in root biomass was
found to be slightly over 50% of the total annual increment of
plant debris.

An effect of climate on prairie productivity was observed in
comparing a northern Iowa prairie with the central Missouri prairie
and also with a southwest Missouri site [40]. Productivity was
greatest in the southern site and least in the northern site. Longer
growing seasons and more rainfall in the southern sites were the
apparent factors responsible. Retarded decomposition at the northern Iowa site resulted in greatest litter accumulation, and litter
accumulation was least in the southwest Missouri site.

B. Estimates of Microbial Biomass

A consideration of microbial growth and carbon turnover requires
a quantitative evaluation of the microbial population, because the
organisms active in decomposition require carbonaceous substrate for
maintenance and the amount of this requirement depends on the microbial biomass present in a given soil. Recent reviews have summarized information on estimating numbers of bacteria by plate
counts [41] and by direct observation [15]. Various methods of estimating fungal activity in soil have also been evaluated [42].

Total biomass in soil is subject to some seasonal variation [43].
The most significant factor controlling both the size of the microbial
population and the evolution of CO_2 was observed to be moisture in a
study of beech forest litter decomposition [44]. In this study,
fungal counts declined immediately when the litter became dry, whereas
bacterial counts showed a delayed depression in numbers.

Alexander [45] estimates that fertile agricultural soils contain 10^8-10^9 bacteria per gram of soil and that the average weight is 1.5×10^{-12} g per cell. This totals 0.15-1.5 g live weight of bacteria per kilogram of soil. He estimates further that soil contains 10-100 m of fungi per gram. Assuming an average diameter of 5 μm and a specific gravity of 1.2, live weight of fungi per kilogram of soil would range from 0.24 to 2.4 g. Living fungi were estimated at near 90 m/g dry soil in a moor profile where the dark colored fungi predominated in nearly all horizons [46]. In a grassland soil, fungal length was at least three times greater than that in the moor [47]. For the moor site, where the investigation was more detailed, bacterial biomass exceeded that for fungi and total biomass of microflora was estimated at between 28 and 197 g wet wt/m^2. Clark and Paul [13] summarized recent estimates of biomass in Matador Grassland soil, reporting average dry weights of 57 g/m^2 per 30 cm for bacteria and actinomycetes combined and 138 g for fungi. In reviewing estimates of bacterial biomass, Clark [48] concluded that soil bacteria probably occupy 0.4-1.0% of the soil volume not occupied by solids or liquids and considered to be theoretically available space for growth.

Classic microscopic estimates of the microbial population may not entirely reflect the quantity of active vegetative microbial tissue in soil. New developments in this area include the bioluminescence technique for determining adenosine triphosphate (ATP) as an index of microbial biomass. Microbial biomass in ocean profiles has been successfully estimated by determining ATP [49]. The assay for ATP in 25 soils gave a range of values somewhat narrower than but characteristic of that for microbial counts [50]. It was concluded that the method is capable of the sensitive detection of ATP in a variety of soil types and that with further development it may prove of value in determining the biological activity or biomass in soils. In a study of ATP in lake sediments, Lee et al. [51] presented a theoretical consideration along with some prelim-

inary experimental results that indicated that the method should
be equally applicable to soils.

V. SUBSTRATE CONTROL OF MICROBIAL GROWTH

A. Microbial Growth Kinetics

Apparatus for the continuous culture of bacteria, such as the
chemostat, provides a laboratory model for studying growth of microorganisms at steady state. A steady-state population in a chemostat may serve as an experimental scale model of natural populations
[19]. A discussion by Herbert et al. [52] of the mathematical
theory of continuous culture and the application of the theory to
experimental data laid the foundation for recent investigations
using the kinetic approach to study microbial growth in natural
habitats. A hyperbolic equation describing microbial growth
kinetics [53] relates the rate of growth per unit of organism to
the concentration of substrate.

Kinetic studies dealing with the rate of growth of aquatic
microorganisms have recently been reviewed by Strickland [54]. Some
investigations suggest concepts that may be relevant to, or methods
that may be adapted for, the study of microbial growth in soil.
Jannasch [55] used the chemostat to study growth responses of
heterotrophic marine bacteria to extremely low concentrations of
simple organic substrates. Threshold concentrations of substrate
were noted below which the organisms were unable to grow. Two
types of organisms were distinguished, one adapted to the marine
environment by its ability to grow at low substrate concentration
and one inactive in seawater but surviving under natural conditions.
He tentatively concluded that species of high growth efficiency at
low substrate concentration would escape detection and isolation
by the usual techniques of batch culture enrichment or by plating.
Conversely, the species isolated from seawater on nutrient agar
of usual strength are probably not representative of active marine
microbial flora. A parallel may exist in soil, where organisms

that feed on available substrate at low concentration and play a
primary role in carbon mineralization may not be those isolated by
the usual techniques. The explanation that Jannasch proposed for
the minimum threshold concentrations that he observed for marine
bacteria was that a positve-feedback mechanism exists that depends
on population density and on metabolic activity.

Recent chemostat design permits experimental testing of the
kinetic equations reformulated to include feedback [56] and this
new design makes possible laboratory studies that more precisely
model a variety of microbial processes. Feedback permits the ac-
celeration of microbial processes where normally the substrate concentration is fixed at a low level or where an inhibitory product
is formed at a concentration proportional to substrate concentration.
A microbial process in soil that conforms to the latter is the oxidation of ammonia to nitrite by *Nitrosomonas*. Processes for treating
of waste waters in which there is recycling of sludge involve cell
feedback and model equations defining growth in the system have been
developed [57]. The kinetic characterization of the growth of microbial populations obtained from a municipal treatment plant has
recently been reported [58]. The maximum rate of growth per unit of
organism for the heterogeneous populations growing on sewage was in
general agreement with the same parameter obtained using glucose or
several other simple organic compounds as carbon sources. Kinetic
parameters for soil populations grown on a variety of substrates,
including soil humus, could be similarly obtained.

Wright and Hobbie [59] used kinetic equations in a study of
substrate utilization by bacteria and algae in aquatic ecosystems.
Methods employing ^{14}C-labeled organic compounds were used to
measure the uptake of organic solutes by planktonic microorganisms.
Turnover time for the substrate, derived from uptake kinetics, indicated an algal uptake of glucose and acetate that amounted to
less than 10% of the bacterial uptake, even though the algal biomass
was orders of magnitude greater than the bacterial. The bacteria
appeared capable of continuous growth at the low levels of substrate

at concentrations too low for heterotrophic growth of algae. Jannasch [60], in estimating growth rates in natural waters, computed specific growth rates for bacteria as low as 0.005 per hour, corresponding to a generation time of 200 hr. He pointed out that growth in natural water is limited primarily by low concentrations of carbon and energy source. Part of the bacterial population may be present in resting and dormant stages but he states that it is not inconceivable that continuous growth occurs at extremely slow rates.

The kinetic approach has been used to measure mineralization of simple organic substrates by microorganism communities in the sediment of a shallow lake [61]. Activity in the sediment was compared with that in overlying lake water by determining turnover time for glucose and acetate and by calculating the maximum velocity of mineralization of the substrate by the natural microflora. The sediment was found capable of mineralizing 24 times as much glucose as an equal volume of overlying lake water. The methodology utilizing uptake kinetics may be equally appropriate to the study of mineralization rates for natural microflora in the various horizons of a soil profile.

B. Equation for Growth of Soil Microorganisms

Living microorganisms require energy both for maintenance and for growth. The maintenance energy for a bacterial cell, defined by Marr et al. [62] as the amount of energy required for purposes not directly coupled with growth, was shown experimentally in continuous culture systems to be independent of the rate of growth. The theory that they expound describes the relationship between substrate consumption and bacterial growth using the equation

$$dx/dt + ax = Y \, ds/dt \qquad\qquad (1)$$

in which x is the concentration of cells; s is the concentration of substrate, t is the time, Y is the yield coefficient or weight of cells produced per unit weight of substrate, and a is the specific maintenance rate and has the dimension of reciprocal time.

The application of this equation to soil systems may proceed on the assumption that a reasonable value for a would be $0.001 \, hr^{-1}$ [17, 43]. An assumed value for Y of 0.35 is in agreement with values [24] suggested previously in Section III, D.

C. Microbial Cell Turnover

Babiuk and Paul [43] made direct observations of bacteria and actinomycete hyphal bits using a fluorescent staining technique in a grassland soil at monthly intervals from April through October. The results were used to calculate microbial biomass after applying appropriate adjustments for moisture content and density. Biomass in the 0 to 30-cm profile ranged from a low in April of 31.9 to a high in June of $76.2 \, g/m^2$. It was estimated that substrate available to the organisms including above-ground plant parts, roots, and exudates would total $500 \, g/m^2/year$. By substituting these data in Eq. (1) they concluded that the microorganisms could reproduce only a few times a year because over half of the available energy would be required for cell maintenance. The large weight of bacterial material estimated from the direct observation technique is plausible only if most of the bacteria in the soil at any time are in an inactive state.

Estimates of fungal biomass in the same soil show it to average more than double the biomass for bacteria and actinomycetes [13]. The total microfloral biomass averaging $200 \, g/m^2$ per 30 cm must, therefore, be considered in the utilization of the available substrate. On the basis of the above investigation, the bulk of the annual substrate is needed for maintenance, leaving very little for microflora that are reproducing.

Babiuk and Paul [43] noted that plate counts yielded smaller estimates of bacterial biomass than the direct observation technique. If only bacterial biomass estimated by plate counts were

used in Eq. (1), a division rate for bacteria more closely in accord
to that observed under laboratory conditions would be indicated.
They concluded that direct microscopy may provide an appropriate
index of total biomass but that the bacterial plate count is a
better estimate of metabolizing cells in the soil.

Gray and Williams [17], using microbial biomass and annual
litter production data for a forest soil, made similar estimates of
required maintenance energy for soil microorganisms using Eq. (1)
and the previously assumed values for a and Y. For a bacterial biomass of 7.3 kg/ha they computed an annual requirement for maintenance of 184 kg of substrate. Total substrate (annual litter production) was 7640 kg/ha leaving 7456 kg available to support microbial growth. Assuming total utilization of this substrate for
bacterial growth the number of generations per year were calculated
from the equation

$$Y(S + xR) = xR \qquad\qquad (2)$$

in which Y is the yield coefficient, S is the substrate available
for growth, x is the concentration of cells in grams (at steady
state), and R is the number of divisions per year. R was found to
be 543 divisions per year; that is, the bacteria divide once every
16 hr.

Substrate utilization exclusively by bacteria is highly improbable because the microbial biomass in the soil was predominantly
fungal. Applying the bacterial maintenance equation to the total
fungal biomass of 454 kg/ha indicated a requirement of 11,363 kg of
substrate per hectare per year for fungal maintenance. This value
exceeds that of total substrate available and allows no substrate
for microbial growth. The authors therefore conclude that the estimates must be in error for either fungal biomass or annual litter
production, that the maintenance constant and yield coefficient are
incorrect, or that the system is losing energy and is not being
maintained at equilibrium.

VI. MICROBIAL RESPIRATION AND CARBON TURNOVER

A. Rate of CO2 Evolution from Soil

The determination of microbial respiration in soil from CO_2 evolution data is a useful index of microbial activity, but in natural systems it is confounded by root respiration. Wiant [63] evaluated microbial respiration in forest soils and concluded that carbon in annual leaf fall for an average forest would be sufficient to yield 653 g CO_2 per square meter per year and would not be expected to exceed 0.3 $g/m^2/hr$. Much higher annual estimates have been obtained when CO_2 evolution is measured for a short period of time during the season of optimum decay and these probably include appreciable root respiration [64].

Kucera and Kirkham [65] studied CO_2 evolution in a mid-Missouri tall-grass prairie and reported an annual total of 1675 g CO_2 per square meter. They estimated that 60% of the total soil respiration or 1005 g CO_2 per square meter per year was caused by microbial processes. Weekly averages showed that respiration ranged from zero during the coldest period to 450 mg CO_2 per square meter per hour during summer. The 60% factor would suggest a maximum microbial respiration rate of 0.27 $g/m^2/hr$. A study of soil respiration for a dry land meadow showed a lower rate of CO_2 evolution from the soil averaging 0.15 $g/m^2/hr$ [66]. Seasonal variations have also been reported for soil under forest [67, 68]. Pot experiments, in contrast to the results from field studies, indicate that the major portion of soil respiration is caused by plant roots [69].

B. Microbial Turnover Estimated from CO2 Evolution

Relationships between microbial numbers and respiration have been reported in a few instances [70-72]. Decomposition of different kinds of litter, as indicated by loss of weight and by evolution of CO_2, was positively correlated with annual mean bacterial and fungal counts. The correlation reported by Gray and Wallace [70]

between bacterial numbers and CO_2 evolution was used to compute a
rate of CO_2 evolution equal to 0.033 mg C per hour per milligram of
bacteria [17]. It was assumed that the evolved CO_2 represented 65%
of the metabolized carbon and the remaining 35% contributed to an
increase in biomass. Because a typical bacterial cell has 50%
carbon, the weight of bacterial cells being produced would equal
two times the weight of carbon assimilated into new cells. Therefore, the increase in biomass would approximate 0.035 mg/hr/mg of
bacteria. From this rate they estimated that bacterial biomass
would double in 28½ hr. Having previously assumed that CO_2 output
measurements may be overestimates and that plate counts of bacteria
are generally underestimates, they concluded that actual growth in
the soil is slower than this estimate.

Clark [48] utilized measurements taken from the literature to
compute CO_2 production relative to bacterial numbers and biomass
for cultures in different phases of growth. For broth cultures in
the stationary growth phase, the production of CO_2 during 24 hr was
estimated to be approximately equivalent to dry cell weight. The
rate used by Gray and Williams [17] to calculate turnover of bacterial biomass in soil, noted above, is about twice that calculated
for bacteria in the stationary phase. A somewhat slower rate for
soil bacteria is suggested by Clark, who estimates production in
1 day at 0.1 mg CO_2 per 10^9 bacteria. He points out that a population of soil bacteria estimated by direct microscopy, all producing
CO_2 at the estimated rate, would considerably exceed observed rates
of CO_2 evolution in the field. The discrepancy becomes more serious
when consideration is given to other soil microorganisms. Clark,
therefore concludes that a high proportion of cells observed microscopically in soil must be in a resting or dormant condition,
with their respiratory activity at an exceedingly low level. Plate
count estimates of the population in field soils are often only 1
or 2% of estimates by direct counting but appear more compatible
with observed rate of CO_2 production in field soils and may more
correctly enumerate the viable and active bacteria of the soil.

Assuming that the evolution of 0.033 mg C per hour as CO_2 results from 1 mg of bacteria, soil respiration rates measured in the laboratory equalling 0.1 mg C per hour per 100 g soil [73] would indicate a bacterial biomass of 3 mg per 100 g of soil. Active fungal hyphae in the soil also may be contributing to the measured respiration. However, fungi were relatively inconsequential when these soils, which were essentially void of plant residues, were incubated in the laboratory, and the calculation interpreted total activity in terms of bacterial biomass. The estimate is reasonable and would correspond to a plate count estimate of 2×10^7 cells per gram of soil. Biomass must increase rapidly immediately after amending the soil with a substrate, such as glucose, because there is a marked increase in rate of CO_2 evolution. Rates of 10-40 times that for unamended soil that develop within 24-48 hr indicate a corresponding exponential increase in microbial biomass. Under these conditions generation times for bacteria must be similar to those calculated for bacteria grown in culture.

One of the soils studied by Chahal and Wagner [73] had been amply fertilized and the fact that CO_2 evolution was extremely rapid following the amendment of this soil with glucose suggested conditions conducive to optimum microbial growth. Mineral nutrient supply was adequate and soil moisture and aeration had been adjusted to provide an optimum environment. It may be appropriate, therefore, to employ the yield constant of 0.118 g bacteria per kilocalorie of energy suggested by Payne [27] to estimate maximum biomass in this soil. A biomass of 100 mg would be optimum for the 250 mg of glucose added to 100 g of the soil. Biomass active in decomposition of soil humus would be negligible. According to Clark [48], a stationary population corresponding to 100 mg of bacterial biomass would evolve 4-5 mg CO_2 per hour. The maximum rate recorded following addition of the glucose was 15 mg CO_2 per hour. The rate exceeds that for a stationary population but falls somewhat short of rates corresponding to the logarithmic growth phase of bacteria in broth cultures [48]. The maximum rate of CO_2

evolution occurred prior to the time that 40% of the substrate carbon had been evolved as CO_2 and, therefore, prior to when utilization of added glucose could be assumed complete [27]. Probably the
maximum CO_2 evolution rate resulted from less than 100 mg of biomass and in that case the rate, presumably reflecting logarithmic
growth in soil, would suggest an efficiency equal to that for logarithmic growth in pure culture.

Under prolonged incubation in the laboratory, the soils studied
by Chahal and Wagner [73] showed a constant rate of CO_2 evolution
from soil organic matter for a period of 1-3 months. This rate
approximated 0.03 mg C/hr/100 g soil and, according to the afore-
mentioned correlation between respiration and biomass [17], suggests
a biomass of 1 mg. If the substrate utilized was 50% carbon and
65% of it was liberated as CO_2, the respiration rate for a period
of 1-3 months indicates a utilization of 0.09 mg of substrate per
hour. Using these values and the 0.35 and 0.001 estimates for yield
and maintenance in Eq. (1) and solving for dx/dt, a value of 0.03
mg/hr is obtained for microbial growth rate. This rate would suggest a turnover for the 1 mg of microbial biomass every 33 hr. The
decomposition of soil organic matter during the 3-month period was
5% and nearly double the annual amount under field conditions.
Therefore, turnover time calculated from data for the laboratory
incubation of soil is assumed to indicate more rapid turnover than
is expected under field conditions.

C. Rate of O_2 Uptake in Soil

On a molar basis, O_2 consumed in aerobic respiration is about
equivalent to CO_2 liberated but relatively few measurements have
been made of O_2 uptake on soil in situ [74]. Most studies in the
field report CO_2 evolution rather than O_2 uptake. Estimates involving field soils have recently been reported in which samples
are brought into the laboratory and special techniques are used to
measure O_2 consumption [75, 76]. These studies suggest a microbial
uptake averaging 10 ml O_2 per 100 g soil per day, but admittedly

the rates are higher than those expected under actual field conditions. For submerged soils, rates equal to 2% of the above have
been reported [77]. Maximum microbial respiration in the field is
probably equivalent to laboratory measurements made on soil samples
at low levels of moisture where uptake amounts to 1 mg O_2 per kilogram soil per hour [78]. A similar rate has been reported for soil
incubated in the laboratory but not until after the lapse of an
initial period of higher respiratory activity extending beyond 6
weeks [79]. A maximum evolution in the field of 0.3 g CO_2 per
square meter per hour [63, 65] cited previously is equivalent to an
O_2 consumption of 0.8 mg/kg soil per hour.

D. Microbial Turnover Estimated from O_2 Uptake

Investigations dealing with the yield of ATP per mole of energy
source dissimilated by microorganisms have been evaluated by Veld-
kampt [80] and allow one to estimate microbial growth from the quantity of O_2 consumed. Under conditions in which glucose was used
almost exclusively as the energy source, *Leuconostoc* dissimilated
glucose via a hexose monophosphate pathway that yielded 1 mole of
ATP per mole of glucose fermented, whereas homofermentative lactic
acid bacteria yielded 2 moles of ATP for each mole of glucose
following the Embden-Meyerhof pathway. Three moles of ATP per mole
of glucose were dissimilated by *Aerobacter aerogenes* grown anaerobically when glucose served as a source of both energy and carbon.
In all cases, a constant of 10 g dry weight of organism produced
per mole of ATP was reported and it duplicated the constant cited
previously [27]. Veldkamp calculated that 3 moles of ATP were
formed per atom of oxygen consumed during aerobic growth of *A. aerogenes* and suggested that this efficiency of oxidative phosphoryla-
tion may be extrapolated to other organisms. Based on these assumptions (30 g cells per 16 g O_2), O_2 uptake at the rate of 0.8 mg/
kg/hr corresponds to a maximum production of microbial tissues of
1.5 mg/hr or a daily production of 36 mg/kg of soil.

The estimate of daily synthesis may be used along with an

estimate, taken from the data of Gray and Williams [17], that 1 kg
of soil contains 200 mg dry weight of bacteria, actinomycetes, and
fungi. The two estimates permit one to calculate microbial turnover.

If total biomass in soil remains constant, the turnover time
for microbial tissue may be calculated using the first-order
Michaelis-Menten equation [81]. The turnover factor k is calculated
as follows

$$k = (2.303/t) \log (x/x - a) \tag{3}$$

in which x is the dry weight of microbial cell material, a is the
production of new material per hour, and t is time. Substituting
the values 0.20 g and 0.0015 g/hr for x and a, respectively, and
with time equal to unity, the calculated value for k is 0.0067.
Turnover time may be calculated from the equation

$$t = 2.303 \log [x/(x - a)]/k \tag{4}$$

The half-life of microbial tissue (when a is 50% of x) would be
103 hr.

VII. TURNOVER OF ^{14}C FROM SUBSTRATES AND MICROBIAL TISSUES ADDED TO SOIL

A. Glucose and Dextran

When substrates tagged with ^{14}C are added to soil, the rate
at which CO_2 is liberated from the substrate as it decomposes can
be monitored independently of the CO_2 arising from decomposition of
native soil organic matter. Soil microflora responsible for the
decomposition of the ^{14}C substrate assimilate a portion of the
label and consequently microbial tissues become labeled with ^{14}C.
If the substrate is a material, such as sugar, that is readily utilized by microorganisms, after a period of time the substrate becomes completely metabolized and further decay to liberate $^{14}CO_2$

is exclusively from microbial products and reflects the turnover of
carbon from the tissues of microorganisms.

Glucose was shown by Chahal and Wagner [73] to decompose very
rapidly in soil and, after a period of several months, about 25% of
the ^{14}C had become incorporated into microbial tissues or soil
organic matter whereas the remainder was liberated as $^{14}CO_2$. During
the latter phases of the incubation, slow release of $^{14}CO_2$ suggested
resistance to rapid decay of microbial tissues and products. A
similar study by Persson [82] showed rapid decomposition of ^{14}C
glucose to complete consumption by soil microorganisms in a few days.
Subsequently, the labeled carbon was mineralized very slowly and
even after 3 years of incubation, 15 to 20% of the ^{14}C remained
in the soil.

During the first few days following addition of glucose, maximum mineralization is reached and microbial growth during this period
is very rapid and probably approaches that for microorganisms arti-
ficially cultured on laboratory media. Subsequent microbial growth
is more typical of that under natural conditions, in which slow CO_2
evolution reflects microbial activity similar to the rate associated
with the decay of soil organic matter. Synthesis of humic materials
known to be highly resistant to decay has no doubt occurred when
the incubation of ^{14}C-glucose is extended over a period of several
months but slow turnover of carbon may also be a reflection of
typical microbial activity in soil.

Accompanying the decomposition of ^{14}C-glucose in soil, inves-
tigators [83, 84] have observed the incorporation of ^{14}C into the
classical fractions of soil organic matter. Humic acid, fulvic
acid, and humin each include a spectrum of components [9]. These
are purely empirical fractions of soil organic matter and are not
very useful in elucidating the biology of the soil [82]. Applying
the classical extraction-fractionation procedure to microbial cells
will also yield materials in each fraction [85]. Therefore, the
distribution of the label among the fractions of soil organic
matter does not necessarily indicate advanced humification of

microbial tissues and it is equally likely that the ^{14}C in any
fraction may have arisen from active or dormant structures of microorganisms.

Mutatkar and Wagner [84] in examining soil organic matter that
became labeled as a result of incubating ^{14}C-glucose in several
soils differing in long term cropping sequence, reported that the
newly formed soil organic matter was less variable in extractability
than that formed some time ago. They concluded that soil organic
matter that became labeled was rather uniform in character for the
several soils studied. Presumably the same general microbial population was stimulated by the addition of glucose. In contrast,
soil organic matter present prior to adding glucose differed in
character among the soils probably as a result of the humification
of diverse substrates and presumably by differing microbial populations associated with each.

A procedure for distinguishing aliphatic from aromatic materials
based on selective combustion of the aliphatics was included in
the investigation [84], even though the reliability of the technique
had never been proved and was generally considered doubtful. The
result indicating a considerable fraction of the ^{14}C in aromatic
structures of microbial origin, therefore, is open to criticism. It
has been established, however, that soil microorganisms are capable
of synthesizing aromatic products and the resemblance of these to
soil organic matter is considered relevant to soil organic matter
formation [86, 87].

The decomposition of polysaccharides in soil may have some
significance to microbial turnover because polysaccharides are
structural elements in microbial cell walls. Soil microorganisms
are known to synthesize a wide variety of polysaccharides and the
biochemistry of these has recently been extensively reviewed [88].
Dextran, a common polysaccharide, was labeled with ^{14}C by Cades
and Wagner [89] and its decomposition in soil was compared to that
of glucose. Carbon dioxide evolution from dextran lagged behind
that from glucose, especially during the first several days of incubation. At the end of 28 days, only 66% of the carbon in dextran

had been liberated as CO_2 compared with 75% of that from glucose.
Near the end of the incubation, the evolution of labeled CO_2 was
similar for the two systems and it was concluded that a portion of
the dextran had become protected from decomposition during the
incubation.

The incorporation of the ^{14}C into sugars other than glucose
at intervals during the incubation was also investigated [89, 90].
Labeled carbon from glucose, during the first few days of incubation,
was rapidly incorporated into other hexoses, but the specific activity of the several sugars changed little over the next month. The
distribution of the label indicated that the microbial population
of the soil synthesized hexoses to a considerably greater extent
than pentoses and deoxyhexoses, the major portion of which might be
derived from plant materials.

An investigation [91] of the tagged carbon in amino compounds
resulting from the decomposition of ^{14}C-glucose accounted for from
17 to 22% of the label in the soil. This label which resulted from
microbial synthesis showed highest specific activity in alanine,
glutamic acid, lysine and glycine, which are known to predominate
in microbial cell walls [92-94]. It was concluded that the persistence of these particular amino acids suggested resistance to decay
of microbial wall tissues in relation to microbial cytoplasm. In
addition, the specific activity for glucosamine was significantly
higher than that for galactosamine and this difference corresponded
to their relative amounts in both bacterial [95] and fungal [96]
cell walls. Glucosamine was found to be the major amino component
of two soils studied; the total ^{14}C associated with it was significantly higher than that for any other amino compound, indicating
a relatively high retention of cell wall material.

B. Microbial Fractions

A comprehensive investigation of the fate of microbial cell
material added to soil was conducted 40 years ago by Jensen [97].
His work showed considerable variation in the rate of decomposition
of microbial cell tissues and demonstrated the presence of resistant
cell fractions. It is significant that among the fungi showing

special resistance to decomposition were *Mycogone nigra* and *Stachy-botrys* sp., both of which are pigmented.

Webley and Jones [98] have recently reviewed transformations of microbial residues added to soil and their resistance to decomposition. Since their review, we have investigated resistance to decomposition of several fungal fractions and this new work will be summarized here. Hurst and Wagner [99] tagged fungal species with ^{14}C and incubated the labeled hypha in soil. Species with high carbon contents and associated dark pigmentation, suggesting the presence of aromatics, fats, and waxes, were more resistant to decay than species that were of low carbon content and of hyaline character and that were presumably richer in carbohydrate. A hyaline fungus, *Aspergillus niger*, and an unnamed melanic fungus isolated from a soil sclerotium were fractionated into cell wall and cytoplasmic components and the fractions were incubated separately in soil. The two cytoplasmic fractions followed similar decomposition patterns. Rapid decomposition occurred a few days after the materials were added to the soil; however, total decomposition for the hyaline cytoplasm was about 20% greater than that for melanic cytoplasm. Decomposition patterns for the two cell wall materials were strikingly different. The hyaline cell wall of *A. niger* decomposed at a relatively high rate for a period extending to 100 days and total decomposition exceeded that for the corresponding cytoplasm. At 167 days, 71% of the *A. niger* cell wall carbon had been liberated as CO_2, an amount approaching that for glucose mineralization in this soil. The melanic cell wall was significantly slower in starting to decay and leveled off very soon in a manner suggesting that only a small portion of this amendment was suitable substrate for soil microorganisms active in decay. At 167 days only 35% of the carbon from melanic cell wall had evolved as CO_2. It was concluded that cell wall substances most likely to be preferentially included in soil organic matter were those of the melanic fungi. Slow growth of these organisms [100] and the indicated resistance to decay suggests that carbon turnover by then would be especially slow.

The widespread occurrence in soil of the melanic fungi is
alluded to in reviews by Burges [100] and Warcup [101]. Witkamp
[102] reported that, in acidic soils rich in humus, the pigmented
hyphae are often present in large amounts. In the consumption of
the substrate debris annually added to soils, the dark walled fungi,
many of which are sexually sterile forms, may be equal to or of
greater importance than the species that grow rapidly and readily
sporulate on synthetic media. Some advances have been made in
developing new techniques for studying the amounts of these dark
hyphae in different soils [103] and a better understanding of the
role of these organisms in soil should be forthcoming.

Fungi that contain no melanin-like material but that apparently
resist microbial lysis under natural conditions also have been
found to occur in soil. One such fungus, *Mortierella parvispora*,
has been reported to possess marked resistance to digestion by microbial enzymes [104]. The walls are found to contain both chitin
and glucan, which are important structural components of other
fungi that do readily lyse. *Mortierella parvispora* differed from
species susceptible to digestion by glucanase and chitinase in that
fucose was found in acid hydrolyzates of wall preparations of the
former.

The development of sclerotia by certain fungi permits their
survival in soil under conditions too adverse or severe for the
ordinary vegetative mycelium [105]. Survival of *Macrophomina
phaseoli* over winter is assumed to result from development of
sclerotia, the mature form of which is darkly pigmented [106]. This
organism was cultured in the laboratory to obtain sclerotia that
were uniformly labeled with ^{14}C and these melanic structures were
fractionated to obtain a cell wall and a cytoplasmic component [107].
Resistance to decomposition in soil of the sclerotial fractions was
studied in comparison with dextran. The cytoplasmic fraction decomposed more slowly than dextran but rapidly in comparison to the
wall fraction. The pattern for cytoplasm was similar to that for

cytoplasm from melanic hyphae reported above. The sclerotial cell wall fraction showed a pronounced lag in rate of CO_2 evolution at the beginning and an exceptionally slow decomposition occurring throughout the 84 days of incubation. At the termination of the incubation only 12% of the carbon from the sclerotial cell wall fraction had been liberated as CO_2. It was concluded that dark colored fungal structures, and particularly sclerotial cell walls, are very resistant to attack by soil microorganisms active in decay.

Using autoradiography, Grossbard [108] investigated the decomposition of ^{14}C-labeled plant fragments and described a unique method of observing microbial growth. Certain microorganisms were shown to be responsible for breakdown of the plant materials by the transfer of ^{14}C into their tissues. Radioactive fungal hyphae were observed in the soil at distances up to 2 cm from the boundary of the added plant material, most frequently in the first 2 to 3 months of incubation when the plant fragments decayed at the fastest rate. The labeled fungi were classified as sterile dark mycelia. Other fungi were present in great profusion but were not radioactive. Later stages of decomposition showed the sterile dark hyphae still present in soil but no longer labeled, whereas some labeled spores and mycelia of actinomycetes were observed.

At the time when leaf fragments had decomposed to the extent that the image density on the autoradiograms had decreased markedly, there appeared concentrated at surfaces of leaf fragments small circular areas of ^{14}C that were identified as fungal sclerotia. Grossbard [108] concludes that microbial activity in soil takes place in small pokcets of organic matter that are not distributed homogenously throughout the soil. Some carbon from decomposing plant residues becomes concentrated within certain microbial structures, which persist in the soil for a long time.

REFERENCES

1.   F. E. Clark, in *The Ecology of Soil Bacteria* (T. R. G. Gray
     and D. Parkinson, Eds.), p. 441, Liverpool Univ. Press,
     Liverpool, 1967.
2.   J. S. Waid, in *The Ecology of Soil Fungi* (D. Parkinson and
     J. S. Waid, Eds.), p. 55, Liverpool Univ. Press, Liver-
     pool, 1960.
3.   M. Alexander, *Microbial Ecology*, Chap. 16, Wiley, New York,
     1971.
4.   B. Bolin, in *The Biosphere*, (A Scientific American Book),
     Chap. 5, Freeman, San Francisco, 1970.
5.   G. E. Hutchinson, in *The Earth as a Planet* (G. P. Kuiper, Ed.),
     Chap. 8, Univ. Chicago Press, 1954.
6.   W. B. Bollen, *Microorganisms and Soil Fertility*, Oregon State
     College Press, Corvallis, 1959.
7.   H. J. Sanders, *Chem. Eng. News*, *44*, March 28, 21A (1966).
8.   C. C. Delwiche, in *Microbiology and Soil Fertility* (C. M.
     Gilmour and O. N. Allen, Eds.), p. 29, Oregon State
     Univ. Press, Corvallis, 1965.
9.   G. T. Felbeck, Jr., in *Soil Biochemistry*, (A. D. McLaren and
     J. Skujins, Eds.), Vol. 2, Chap. 3, Dekker, New York,
     1971.
10. J. C. Anway, E. G. Brittain, H. W. Hunt, G. S. Innis, W. J.
     Parton, C. F. Rodell, and R. H. Sauer, *Grassland Biome
     U.S. I.B.P. Tech. Rept.*, *156*, 1972; and M. W. Hunt,
     Colorado State University, Fort Collins, private com-
     munication, 1973.
11. D. Parkinson, in *Soil Biology* (A. Burges and F. Raw, Eds.),
     Chap. 15, Academic, London, 1967.
12. A. D. Rovira and B. M. McDougall, in *Soil Biochemistry* (A. D.
     McLaren and G. H. Peterson, Eds.), Vol. 1, Ch. 16, Dekker, New
     York, 1967.
13. F. E. Clark and E. A. Paul, *Advan. Agron.*, *22*, 375 (1970).
14. N. A. Krasilnikov, *Soil Microorganisms and Higher Plants*
     (English transl.), Part II, The Israel Program for
     Scientific Translations, 1961.
15. T. R. G. Gray, P. Baxby, I. R. Hill, and M. Goodfellow, in
     The Ecology of Soil Bacteria (T. R. G. Gray and D.
     Parkinson, Eds.), p. 171, Liverpool Univ. Press, Liver-
     pool, 1967.
16. T. R. G. Gray, in *Pesticides in the Soil: Ecology, Degrada-
     tion, and Movement*, p. 36, Michigan State University,
     East Lansing, 1970.
17. T. R. G. Gray and S. T. Williams, *Microbes and Biological
     Productivity*, *(Symp. Soc. Gen. Microbiol.)*, *21*, 255
     (1971).
18. T. R. G. Gray and S. T. Williams, *Soil Micro-organisms*,
     Oliver and Boyd, Edinburgh, 1971.
19. T. D. Brock, *Bacteriol. Rev.*, *35*, 39 (1971).

20. B. V. Perfil'ev and D. R. Gabe, *Capillary Methods of Investigating Micro-organisms*, Oliver and Boyd, Edinburgh, 1969.
21. L. E. Casida, Jr., *Appl. Microbiol.*, *18*, 1065 (1969).
22. H. Katznelson and I. L. Stevenson, *Can. J. Microbiol.*, *2*, 611 (1956).
23. S. A. Waksman, *J. Amer. Soc. Agron.*, *21*, 1 (1929).
24. H. Lees and J. W. Porteous, *Plant Soil*, *2*, 231 (1950).
25. E. Terroine and R. Wurmser, *Bull. Soc. de Chim. Biol.*, *4*, 519 (1922).
26. B. V. Siegel and C. E. Clifton, *J. Bacteriol.*, *60*, 113 (1950).
27. W. J. Payne, *Ann. Rev. Microbiol.*, *24*, 17 (1970).
28. P. S. Vary and M. J. Johnson, *Appl. Microbiol.*, *15*, 1473 (1967).
29. W. T. Blevins and J. J. Perry, *Z. Allg. Mikrobiol.*, *11*, 181 (1971).
30. T. R. Camp, *Water and Its Impurities*, Chap. 9, Reinhold, New York, 1963.
31. A. F. Gaudy, Jr. and Mr. Ramanathan, *Biotech. Bioeng.*, *13*, 113 (1971).
32. T. D. Brock, *Biosciences*, *17*, 166 (1967).
33. A. Burges, in *Soil Biology* (A. Burges and F. Raw, Eds.), Chap. 16, Academic, London, 1967.
34. D. S. Jenkinson in *The Use of Isotopes in Soil Organic Matter Studies* (FAO/IAEA Symposium, 1963), p. 187, Pergamon, Oxford, 1966.
35. H. Jenny, *Factors of Soil Formation*, Chap. 7, McGraw-Hill, New York, 1941.
36. C. L. Kucera, R. C. Dahlman, and M. R. Loelling, *Ecology*, *48*, 536 (1967).
37. A. Macfadyen, in *Methods for the Study of Production and Energy Flow in Soil Communities* (J. Phillipson, Ed.), UNESCO, Blackwell, Oxford, 1970.
38. J. Satchell, in *Production of the World's Forests*, (J. Duvingneaud, Ed.), Brussels, 1970.
39. R. C. Dahlman and C. L. Kucera, *Ecology*, *46*, 84 (1965).
40. M. R. Loelling and C. L. Kucera, *Iowa State J. Sci.*, *39*, 387 (1965).
41. V. Jensen, in *The Ecology of Soil Bacteria* (T. R. G. Gray and D. Parkinson, Eds.), p. 158, Liverpool Univ. Press, Liverpool, 1967.
42. J. H. Warcup, in *The Ecology of Soil Fungi* (D. Parkinson and J. S. Waid, Eds.), p. 3, Liverpool Univ. Press, Liverpool, 1960.
43. L. A. Babiuk and E. A. Paul, *Can. J. Microbiol.*, *16*, 57 (1970).
44. M. E. Dutch and J. D. Stout, *Trans. 9th Int. Cong. Soil Sci.*, Adelaide, *2*, 37 (1968).
45. M. Alexander, *Introduction to Soil Microbiology*, Wiley, New York, 1961.
46. P. M. Latter and J. B. Cragg, *J. Ecol.*, *55*, 465 (1967).
47. P. M. Latter, J. B. Cragg, and O. W. Heal, *J. Ecol.*, *55*, 445 (1967).

48. F. E. Clark, in *Soil Biology* (A. Burges and F. Raw, Eds.), Chap. 2, Academic, London, 1967.
49. O. Holm-Hansen, *Limnol. Oceanogr.*, *14*, 740 (1969).
50. N. H. MacLeod, E. H. Chappelle, and A. M. Crawford, *Nature (London)*, *223*, 267 (1969).
51. C. C. Lee, R. F. Harris, J. D. H. Williams, D. E. Armstrong, and J. K. Syers, *Soil Sci. Soc. Amer. Proc.*, *35*, 82 (1971).
52. D. Herbert, R. Elsworth, and R. C. Telling, *J. Gen. Microbiol.*, *14*, 601 (1956).
53. J. Caperon, *Ecology*, *48*, 715 (1967).
54. J. D. H. Strickland, in *Microbes and Biological Productivity (Symp. Soc. Gen. Microbiol.)*, *21*, 231 (1971).
55. H. W. Jannasch, *Limnol. Oceanogr.*, *12*, 264 (1967).
56. S. J. Pirt and W. M. Kurowski, *J. Gen. Microbiol.*, *63*, 357 (1970).
57. M. Ramanathan and A. F. Gaudy, Jr., *Biotech. Bioeng.*, *13*, 125 (1971).
58. K. M. Peil and A. F. Gaudy, Jr., *Appl. Microbiol.*, *21*, 253 (1971).
59. R. T. Wright and J. E. Hobbie, *Ecology*, *47*, 447 (1966).
60. H. W. Jannasch, *J. Bacteriol.*, *99*, 156 (1969).
61. M. J. Harrison, R. T. Wright, and R. Y. Morita, *Appl. Microbiol.*, *21*, 698 (1971).
62. A. G. Marr, E. H. Nilson, and D. J. Clark, *Ann. N.Y. Acad. Sci.*, *102*, 536 (1962).
63. H. V. Wiant, Jr., *J. Forestry*, *65*, 408 (1967).
64. G. W. Wallis and S. A. Wilde, *Ecology*, *38*, 359 (1957).
65. C. L. Kucera and D. R. Kirkham, *Ecology*, *52*, 912 (1971).
66. M. Losonen, *Oikos*, *19*, 242 (1968).
67. M. Witkamp and J. van der Drift, *Plant Soil*, *15*, 295 (1961).
68. W. A. Reiners, *Ecology*, *49*, 471 (1968).
69. Ye. I. Shilova, *Sov. Soil Sci.*, *5*, 663 (1967).
70. P. H. H. Gray and R. H. Wallace, *Can. J. Microbiol.*, *3*, 191 (1957).
71. M. Witkamp, *Ecology*, *47*, 194 (1966).
72. P. M. Latter, J. B. Cragg, and O. W. Heal, *J. Ecol.*, *55*, 445 (1967).
73. K. S. Chahal and G. H. Wagner, *Soil Sci.*, *100*, 96 (1965).
74. G. Stotzky, in *Methods of Soil Analysis* (C. A. Black, Ed.), p. 1550, American Society of Agronomy, Madison, Wisconsin, 1965.
75. J. W. Birch and M. Melville, *J. Soil Sci.*, *20*, 101 (1969).
76. D. J. Ross, *J. Soil Sci.*, *16*, 73 (1965).
77. R. H. Howeler and D. R. Bouldin, *Soil Sci. Soc. Amer. Proc.*, *35*, 202 (1971).
78. A. J. Rixon, *J. Soil Sci.*, *19*, 56 (1968).
79. S. D. Lyda and G. D. Robinson, *Soil Sci. Soc. Amer. Proc.*, *33*, 92 (1969).

80. H. Veldkamp, in *The Ecology of Soil Bacteria* (T. R. G. Gray
     and D. Parkinson, Eds.), p. 201, Liverpool Univ. Press,
     Liverpool, 1967.
81. L. Michaelis and M. L. Menten, *Biochem. Z.*, *49*, 333 (1913).
82. J. Persson, *Lantbrukshogsk. Ann.*, *34*, 81 (1968).
83. J. Mayaudon and P. Simonart, *Plant Soil*, *9*, 376 (1958).
84. V. K. Mutatkar and G. H. Wagner, *Soil Sci. Soc. Amer. Proc.*,
     31, 66 (1967).
85. J. Mayaudon and P. Simonart, *Ann. Inst. Pasteur*, *105*, 257
     (1963).
86. B. J. Bloomfield and M. Alexander, *J. Bacteriol.*, *93*, 1276
     (1967).
87. K. Haider and J. P. Martin, in *Soil Biochemistry*, (E. A.
     Paul and A. D. McLaren, Eds.), Vol. 3, Chap. 12, Dekker,
     New York, 1972.
88. P. Finch, M. H. B. Hayes and M. Stacey, in *Soil Biochemistry*,
     (A. D. McLaren and J. Skujins, Eds.), Vol. 2, Chap. 9,
     Dekker, New York, 1971.
89. J. M. Oades and G. H. Wagner, *Soil Sci. Soc. Amer. Proc.*, *35*,
     914 (1971).
90. J. M. Oades and G. H. Wagner, *Geoderma*, *4*, 417 (1970).
91. G. H. Wagner and V. K. Mutatkar, *Soil Sci. Soc. Amer. Proc.*,
     32, 683 (1968).
92. M. R. J. Salton, *The Bacterial Cell Wall*, Chap. 4, Elsevier,
     Amsterdam, 1964.
93. E. M. Crook and I. R. Johnston, *Biochem J.*, *83*, 325 (1962).
94. R. Mitchell and N. Sabar, *Can. J. Microbiol.*, *12*, 471 (1966).
95. E. F. Gale, *Synthesis and Organization in the Bacterial Cell*,
     Chap. 1, Wiley, New York, 1959.
96. J. M. Aronson, in *The Fungi an Advanced Treatise I. The
     Fungal Cell* (G. C. Ainsworth and A. S. Sussman, Eds.),
     Chap. 3, Academic, New York, 1965.
97. H. L. Jensen, *J. Agri. Sci.*, *22*, 1 (1932).
98. D. M. Webley and D. Jones, in *Soil Biochemistry* (A. D. McLaren
     and J. Skujins, Eds), Vol. 2, Chap. 15, Dekker, New
     York, 1971.
99. H. M. Hurst and G. H. Wagner, *Soil Sci. Soc. Amer. Proc.*, *33*,
     707 (1969).
100. A. Burges, *Micro-organisms in the Soil*, Chap. 2, Hutchinson,
     London, 1958.
101. J. H. Warcup, in *Soil Biology* (A. Burges and F. Raw, Eds.),
     p. 51, Academic, London, 1967.
102. M. Witkamp, *Mededelingen Inst. Toegepast Biol. Inderzoek
     Natuur*, *46*, 1 (1960).
103. D. P. Nicholas, D. Parkinson, and N. A. Burges, *J. Soil Sci.*,
     16, 258 (1965).
104. R. M. Pengra, M. A. Cole, and M. Alexander, *J. Bacteriol.*,
     97, 1056 (1969).
105. B. B. Townsend and H. J. Willetts, *Brit. Mycol. Soc. Trans.*,
     37, 213 (1954).

106.   T. D. Wyllie and M. F. Brown, *Phytopathology*, *60*, 524 (1970).
107.   G. H. Wagner and H. Rabinska, *Studies about Humus, Humus et Planta V*, Trans. Int. Symp., *1*, 27, Prague (1971).
108.   E. Grossbard, *J. Soil Sci.*, *20*, 38 (1969).

AUTHOR INDEX

Numbers in parenthesis are reference numbers and indicate that an author's work is referred to although his name is not cited in the text. Underlined numbers give the page on which the complete reference is listed.